NEWLY UPDATED 2020 EDITION

Long Island Indicator Service

Companion Reference Guide for Test Indicators

by René Urs Meyer

with content based on the Long Island Indicator Service web site

www.longislandindicator.com

Other repair manuals by René Urs Meyer

Available from our website or through online book sellers world-wide.

BesTest and TesaTast Indicators
Compac Test Indicators
Interapid Test Indicators
Starrett Last Word Indicators
Mitutoyo (pre-2017) Test Indicators

2020 Edition

© 2020 by René Urs Meyer

CONTENTS

Test Indicator Brand Comparison 9
Brown & Sharpe BesTest 24
Girod-Tast: The other Swiss test indicator 37
Compac Swiss Test Indicator 40
Interapid Test Indicator 42
MarTest Test Indicator 66
Mitutoyo Test Indicator 68
Starrett Indicators 84
Tesatast Test Indicator 93
Test Indicator Holders 97
Test Indicator Contact Points 100
Indicator Model Equivalents 104
Calibration of gages: Calibration procedures 116
Test Indicators: What can go wrong? 119
Indicator Crystal Installation Reference Page 124
How to read the graduations on Indicator Dials 131

PREFACE TO THE 2020 EDITION

Everything has changed.

BesTest indicators, the pride of Swiss gages, are now made in China. Compac indicators our best sellers since 1960 are discontinued. Starrett managed to create an impressive catalog. Mitutoyo completely revamped all of their test indicators in 2017. Mahr bought Federal and promises to give Brown & Sharpe some heady competition.

I retired from 40 years of repairing indicators. My brother and his team now run Long Island Indicator Service. I have completely redesigned our web site and, since Spring 2020, I have been living just a stone's throw from Waikiki Beach.

It was time to update the original Companion Reference Guide. I trust that it will still prove useful in many ways.

René Urs Meyer,
Honolulu, Hawai'i

PREFACE TO THE 2017 EDITION

Let me make it clear from the outset that what follows is quite simply a reprint of certain portions of our web site. Specifically, those portions dealing with dial *test indicators*.

This book came about as a result of the many requests we have had for a print version, something that can be kept on the bookshelf—a reference perhaps—or reading material, when sitting in front of the computer screen is not an option.

When we began the project it became stunningly evident that a print version of the complete web site would exceed 3,000 pages. That could amount to an encyclopedia of ten volumes not to mention quite a bit of work on my part.

I also noticed that a straight forward print-out would not make much sense. Web sites are designed in non-linear form. That is, one web page does not lead logically to the next. Each web page contains numerous links which will have the visitor jump all over the site without necessarily being aware that this is a spider web, a maze of information.

In order to keep a direct correlation between the printed chapters in this book and what we call "web pages" on the computer, I have had to make reference to the website's web page numbers.

In the text that follows, any reference to web page numbers refers to the web page numbering on our web site. Thus, when the reader finds a topic of interest, it is an easy matter to go to the computer and locate the same information.

To make this book manageable in size, I have chosen to focus on test indicators alone. Indicator repairs are what we started with in 1959 and specifically the Swiss-made gages.

My father, George, arrived at Idlewild airport only to be told that the job he had been promised fell through. Quite possibly the job never existed and he was lured here under false pretenses. Fortunately the Swiss embassy had just received a job listing for a repairman—a repair technician as he would eventually be called—for the exclusive importer of the Swiss-made gages. George was trained to repair these and was sent back to Switzerland several times to get additional factory training.

He started his own business when it became clear that other brands, American made brands such as Brown & Sharpe, Federal and Starrett, also required servicing. This was soon followed by other foreign brands when, around 1960, a small group of salesmen from Japan started to make the rounds in the New York area, selling gages from a four-page leaflet printed on newsprint. Mitutoyo is now undoubtedly the biggest player in the field.

George eventually turned this home business into Long Island Indicator Service and when I joined in 1980 we set about to expand our services and sales. Thanks to the Internet we survived the exodus of the great aeronautic manufacturers which once populated Long Island. Our business suddenly became national and our website became a helpful source of information on precision measuring gages.

The website is a continual work in progress. New information gets added about half a dozen times *each day*. I started to do this simply to create a reference source for our own use. How else could we remember all the information that was unearthed?

Looking at the Table of Contents, you will see that the focus remains principally on the quality indicators made by established manufacturers in Switzerland, Germany, Japan and the USA.

I feel justified in focusing on what we know best. I recently calculated that I have personally repaired over 35,000 indicators in the years that I have been working. My father easily matched that number along with tens of thousands of dial calipers and micrometers which he and my brother have repaired. What I learned during all these years has all been put into this repository: www.longislandindicator.com

Here you will find charts of manufacturers' specifications and spare parts lists, critical evaluations of brands and models, instructions for cleaning, servicing and calibrating indicators and hundreds of web pages of information which often exceed what manufacturers' catalogs provide. In fact, we have had manufacturers refer to our website to learn things about their own products!

Not always in our favor, however. We have been threatened with a lawyer's visit when we were overly critical; we have been asked to remove proprietary information that the manufacturers had rather remained secret; and, we have lost a major distributorship when we dished some of the very products we were supposed to be selling.

Yet, we have acted out of a sense of responsibility. The desire is to provide honest, straightforward information; to guide the user away from cheap, poorly built gages; to add to this country's ability to produce quality products by using quality gaging suited to the task. We want American manufacturing to be a premier example of what is possible and we feel we are an accountable partner in this endeavor.

René Urs Meyer
Long Island, New York

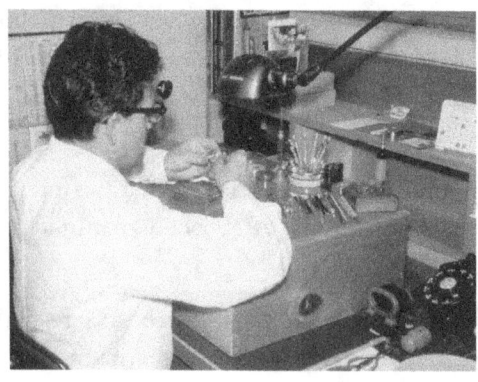

My father at his repair desk ca. 1964

TEST INDICATOR BRAND COMPARISON

In Brief: the best dial test indicators are Swiss made. Take that with a heaping spoonful of salt, however. Even "Swiss" indicators will contain Chinese elements and they may even be assembled in China. As a matter of conjecture, we can suspect that ALL manufacturers now resort to the aid of China in making their gages.

Here's the odd part: The worst dial test indicators are Chinese, and - sorry to say - American. We'll let you figure out the manufacturers in question.

Don't have time to do a lot of reading and research and are willing to trust us for the best choices regardless of their country of origin? Here they are (without any reservations):

- Single revolution (.0001" or .0005"): BesTest
- Two revolutions (.0001" or .0005"): Interapid

NOTES ON MANUFACTURERS

Accupro vanity dial appears on indicators made in China and Germany. The Chinese indicators are worthless and can not be repaired. The German indicators are the Puppitast series made by Mahr-Federal.

- Repairs: see Mahr-Federal listing, below
- Sales: catalogs
- Parts: from Mahr-Federal for the German made models only
- Information: see Mahr listing, below

Alina (Switzerland) indicators were made by Compac until the mid-1960's. They are no longer available and spare parts are exhausted. The Alina Model 88 indicator was a superior version of the American-made Last Word indicator.

Baker (China) indicators are cheap throw-aways for which parts are not available. We have been told that they are somewhat longer lived than other Chinese brands. Replacement contact points are not available but Compac points will fit, in a pinch.

Baty (Swiss) indicators are a vanity dial on Girod-Tast indicators. (see below)

Bestest (Switzerland) has become America's favorite and there are good reasons. They're among the very best available; a great value for the money. Excellent repeatability and quick response make them desirable. If there's a drawback, it's that they're prone to damage because of the light construction. Available in black or white, horizontal, vertical or parallel. Except for the name on the dial and the accessories in the kit, they are identical to Tesatast. Distributed in the US by Brown & Sharpe.

- Repairs: Long Island Indicator Service (an authorized repair shop)
- Sales & Information: Long Island Indicator Service
- Parts: Long Island Indicator Service

China indicators have now flooded the market, being sold under many different brand names such as Türlen (to make them sound German) or even Amazon (to make them sound like they're produced by the internet giant). They cost a fraction of the European or Japanese-made indicators and must be treated as throw-away gages.

- Repairs: not possible
- Sales: discount catalogs everywhere
- Parts: not available

CDI (Chicago) test indicators are identical to Compac (Switzerland). These were made for CDI in the 1980's. CDI no longer sells them but you can buy the Compac replacements.

Compac (Switzerland) a long standing brand originally sold under the name PARVUS, later as Alina, Lufkin and SPI. Because of their high cost, they lost popularity and were slowly phased out starting in 2001. The line was completely discontinued in 2020. Repairs may still be possible at Long Island Indicator Service

Craftsman indicators are sold by Sears but are often times made in the UK in which case they are identical to Verdict indicators. They're not very good (in fact, they're downright awful) but do offer the "pear shaped" contact point which makes them look quite medieval and eliminates the cosine error, in theory.

EMS (Germany) no information available at this time.

Federal Gage made the worst test indicator you could get stuck with. Blobs of solder were used to hold it together. Mercifully these have been discontinued. The last models named Testmaster were made by TESA in Switzerland and they are identical to Bestest indicators (see above). These are no longer available from Federal, but you can still buy the Bestest equivalent. The newest indicators are called MarTest (see Mahr-Federal). Repairs: no longer possible

Fowler once relied heavily on English imports such as Verdict indicators. These were about as good as English weather. Nowadays they rely more heavily on Swiss made gages but also offer lookalikes in their effort to remain competitive. Beware of wolves in sheep's clothing: they offer a pathetic imitation of the Bestest indicator and an Interapid look-alike is made in China and sold under the name Xtest. The best mechanical test indicator which Fowler offers is the Swiss made Girod-Tast. Fowler recently introduced Ultra-Tast indicators which are made by Kafer in Germany. It is a respectable manufacturer but has been known to outsource to China. Repairs and spare parts may be hard to come by. A five-year warranty sounds great but it is against manufacturing defects. Any defects would be noticed within the first few weeks of use and not likely after four-and-a-half years. When shopping Fowler, if it doesn't say "Swiss Made" or "German made": Buyer beware.

- Repairs: limited by parts availability. Ask before sending these.
- Sales: Fowler distributors and online
- Parts: check with Fowler
- Information: Fowler web site

Gem (USA) makes an inferior version of the popular Starrett Last Word Indicator. This would be fine if they were cheaper. There is an odd variation, however: one model has two dial faces, one on each side. This comes in handy in some applications. Some of the newest models have replaceable dove tails. Gem also manufactures a line of indicator clamps and holders.

- Repairs: not worthwhile
- Sales: discount catalogs
- Parts: from the manufacturer or through a dealer
- Information: see their catalog at www.thomasregister.com

Girod-Tast (Switzerland) is similar to the old style (1970's) Bestest indicator with some improvements to make them sturdier. In the USA these are sold by Fowler. In Switzerland they are also sold with the name SISO-Tast. If you've ever wanted a Bestest indicator with multiple revolutions, Girod offers several models with extended ranges. If they have a drawback, it's that the contact point is adjustable. You'd think this were an advantage, but for most people it's a nuisance. On the Bestest you simply unscrew the old, screw in the new. On the Girod-Tast you have to adjust the new point so that the indicator is in calibration.

- Repairs: limited by parts availability. Ask before sending these.
- Sales: Fowler distributors nationwide
- Parts: hard to find

H.G. Jensen (Waltham, Massachusetts) is an exceptionally basic relic of early test indicator design. If you find one of these, keep it for historical purposes.

Hughes Aircraft Company had a patented modification of the Swiss Compac indicator in the 1980s. This allowed it to measure force. It appears that the only modifications to the 214GA model were a stronger return spring, and a custom counter-clockwise dial without a revolution counter.

Interapid (Switzerland) is the gem of all test indicators. These have the distinctive slanted dial which the other manufacturers have only just begun to copy. Correct readings are obtained when the contact angle is 12°. Undoubtedly this has its advantages as long as the user remembers to take it into account. The revolution counter hand does not have any numbers associated with it. There are just a couple of tick marks showing you that you've gone around once or twice. Dials are balanced and the right side of the dial has a thin black line which will help you determine plus or minus in a mirror set-up. A 4 mm diameter holding stem is permanently attached to the far end of the indicator. Models with 2.8" long contact points tend to have a slower response and should probably only be used to measure .001" (Note: beware of cheap Interapid look-alike ripoffs now being offered in catalogs. They're made in China and they're junk. Insist on the real thing.)

- Repairs: Long Island Indicator Service
- Sales & Information: www.longislandindicator.com
- Parts: Long Island Indicator Service

Johnson Gage test indicators of the 1950's and 1960's were made by Compac, Geneva. They were the same as those sold under the Alina brand name (see Alina, above). They are obviously long obsolete.

Kafer (Germany) (also spelled Käfer and Kaefer) manufactures a complete line of test indicators with one revolution. These are beautifully crafted and come in a box with a clear lid, so you can easily see what you're taking off the shelf. Alas, they do not have identifying serial numbers. An excellent alternative to Swiss-made indicators but model styles are limited. Parts are available from Germany but rarely does anyone stock them.

Kurt (USA) although located in Minneapolis, these are generic made-in-China imports. They're cheap throwaways although Kurt claims they are of better quality than other Chinese indicators. This is possible but we can not verify it.

- Repairs: never economically feasible
- Sales: Kurt distributors nationwide and some catalog houses
- Parts: Kurt claims to carry parts

Last Word (USA) Starrett makes this stalwart and ubiquitous test indicator without resorting to toothed gears. Although usually accurate we've seen enough of them that compare poorly with the better built, gear driven indicators to warrant skepticism. The body on older models, being made of iron, rusts easily and will become magnetic (and sticky as a result). Newer models are black anodized.

- Repairs: Starrett's service department and independent repair shops
- Sales & Information: www.longislandindicator.com
- Parts: available directly from the manufacturer

Lufkin never manufactured any of their own indicators. In the 1960's they had a vanity dial on the Alina indicator. These tended to have model numbers such as V60X. It was never clear how they managed to usurp Alina's exclusive rights to these gages and that may have been the reason the line was finally dropped. Repairs are no longer possible due to the obsolete parts.

MarTest possibly manufactured in Czech Republic by Mahr but some gages do not show a country of origin which is a pretty sure sign that they come from China. These are modeled after the Mitutoyo design but the bezel is made of metal and less likely to be damaged. These test indicators have the contact point length conveniently inscribed on the side of the case. We find that the .0001" indicators may be too sensitive for some users. The contact point swivels very easily and this can cause problems with repeatability. The bezel turns on a rubber o-ring and this has sometimes dried out on us creating much too much friction for comfort.

- Repairs: independent repair shops
- Sales & Information: www.longislandindicator.com
- Parts: can be bought directly from the manufacturer

Mahr (Puppitast) manufactured in Germany, part of the pre-Mahr-Federal conglomeration. These are structurally similar to Bestest, Tesatast and Girod-Tast indicators. The handsome bodies are somewhat sturdier and have textured sides which might, under some circumstances, keep them from slipping out of your hands. The crystal can rather easily be replaced without tools and this is an advantage over Bestest and Tesatast. Discontinued.

Mercer manufactured in Switzerland and are identical to Compac indicators. All models are discontinued. Previous models were made in England. They are also discontinued.

MHC Industrial Supply made in China. Whenever the country of origin is not printed on the indicator dial, you can be assured it is Chinese. For some reason they can get away with that.

Mitutoyo new models, completely redesigned in 2017, are manufactured in Japan. Some models are available with optically scannable serial numbers on the dial face. The proper contact point length is now shown on the dial of each indicator. The new "pocket" models 513-512 and 513-518 are a major improvement in design and construction over the old models and can be recommended. The multi-revolution slanted dials which mimic the Swiss Interapid indicator have one significant difference: they are accurate when the contact point is used at an angle of 0°. This could be a source of confusion—and error—in a shop which uses both brands.

- Repairs: Long Island Indicator Service
- Sales & Information: www.longislandindicator.com
- Parts: www.longislandindicator.com

Mueller old models were made in England. Repairs are no longer possible.

Nork indicators were manufactured in Manhattan of all places, by General Howe Mfg Co., Inc. They're a dreadful imitation of the Starrett Last Word indicator although they did have a much more functional reversing lever. Repairs: not possible

Parvus indicators were manufactured in Switzerland during the 1940-50's and sold in the US with the Alina name on the dial. These were later transformed into the Compac models. You may see the word Parvus stamped on some of the old bodies. Long obsolete (1950's), there are no parts or repair service available. Repairs: not possible.

Peacock (Pic-Test) manufactured in Japan. This is a meager entry in the test indicator market, designed along the lines of the old model Bestest. Comparison ends there, however. Calibration often has to be fudged by changing the contact point angle on the .0001" model. Newer models contain plastic gears. They are available from some catalog houses but parts are generally unavailable.

- Sales: various catalogs
- Parts: generally unavailable
- Information: 1-408-871-7700
-

Scherr Tumico (S-T Industries) carries some test indicators made in England and others from China.

- Sales: various catalogs
- Parts: generally unavailable
- Information: see their web site

Shars generic indicator made in China (see China, above)

Sisotast manufactured in Switzerland. This is a vanity dial for the Girod-Tast indicator. The indicators are identical with the exception of the dial.

- Information: see Girod-Tast (above)

SPI (China) manufactured for SPI. These are generally the same Chinese indicators you can buy under any number of other "brand" names. The cheap price gives them away. SPI stands for Swiss Precision Instruments. Don't let this fool you. These are not Swiss and their precision is short lived. (SPI used to offer genuine Swiss indicators with the SPI name. They were made by Compac and you can still get them. See Compac above.)

- Repairs: not possible
- Sales: any SPI distributor
- Parts: not available. Never have been, never will be.
- Information: you read it here

Spot-On made in England, looks for all the world like an old Verdict indicator. These are obsolete and may date back to the 1940-50's if not earlier. Starrett Last Word would be an acceptable replacement if you like this style.

Standard Check-Master manufactured in Poughkeepsie, a long time ago. This indicator was like the Federal TestMaster design only much better. It was elegant and beautiful in comparison. Parts and service are no longer available on this long obsolete item.

Starrett (USA) would like us to believe that they are products of the USA. Recently released models 3808 and 3809 do not claim to be made in USA and rumor has it that these are of Chinese origin. Models 708 and 709 are "American Made". We have found these to be surprisingly accurate. Construction-wise, none of the Starrett test indicators are in the same league as their European made counterparts.

- Repairs: the manufacturer or independent repair shops
- Sales: www.longislandindicator.com
- Parts: can be bought directly from Starrett or a distributor

Teclock (Japan) You can often buy European-made models for less, and you'll get better quality. Spare parts are not commonly available. These indicators are heftier but feature an inferior execution of the Bestest-style mechanism. The newest models seem to come with plastic bezels. When the crystals are scratched, or the bezel breaks (it will) you won't be able to replace them.

- Repairs: generally not possible
- Sales: various catalogs
- Parts: not available

Tesatast (Switzerland but models since 2018 may have Chinese origins) manufactured by TESA are identical to Bestest with all the same good features. The accessories that come with the indicator are different. We have all parts in stock.

- Repairs: Long Island Indicator Service
- Sales & Information: www.longislandindicator.com

Testmaster (USA) an indicator made by Federal Gage and discontinued, mercifully, in the 1970's. This was one of the worst designs and executions of all time. Blobs of solder were used to keep the return spring in place. Unbelievable.

- Parts: obsolete, thank goodness

Türlen an inexpensive generic test indicator with a fancy German-sounding name made in China by the looks of it.

- Repairs: not worth it
- Sales: widely available in discount catalogs
- Parts: not available

Valueline an inexpensive generic test indicator sold by Brown & Sharpe. Made in China by the looks of it

- Repairs: parts not available
- Sales: www.longislandindicator.com

XTest (China) manufactured for Fowler as a rip-off on the high-quality Interapid indicator. They look so much alike in the advertisements that many people are fooled into thinking they're getting a terrific deal on the Swiss indicator. You get what you pay for. In this case, a pathetic imitation.

- Repairs: not possible
- Sales: you're on your own
- Parts: Fowler claims to have parts in stock
- Information: you read it here

Test Indicator Score Card

BRAND	1	2	3	4	5	6	7	8	9	10	11	12	13	14	15	16	17	18
BESTEST	C	A	A	A	A	A	A	A	A	B	A	A	A	A	A	B	C	A
TESATAST	C	A	A	A	A	A	A	A	A	B	A	A	A	A	A	B	C	A
GENERIC	C	A	C	C	C	A	A	A	A	B	A	C	A	A	B	B	C	A
GEM	B	B	B	C	A	A	B	A	A	A	A	C	C	A	B	B	C	C
GIRODTAST	C	B	B	A	A	A	A	A	A	B	A	A	A	A	A	A	C	A
INTERAPID	C	A	A	A	A	B	B	B	B	A	A	A	C	A	B	C	C
KAFER	A	B	C	A	A	A	A	A	A	B	A	A	A	A	A	B	B	C
PUPPITAST	C	A	C	A	A	A	A	B	A	B	A	A	A	C	B	C	?
MARTEST	C	A	B	A	A	C	A	A	A	B	A	A	A	C	A	C	A
MITUTOYO 'POCKET'	C	A	A	A	A	A	A	A	C	A	A	C	A	C	B	C	B
MITUTOYO	A	A	A	C	A	A	B	A	A	B	C	B	A	A	C	B	C	B
LAST WORD	B	B	B	A	C	A	B	C	A	C	A	C	C	B	B	B	C	C
STARRETT 708 & 709	C	B	B	A	A	A	B	B	B	B	B	C	A	A	C	A	B	C
TECLOCK	C	A	C	A	A	A	B	B	A	A	B	B	B	A	C	B	C	?
PEACOCK	C	A	C	A	C	A	A	B	A	B	C	B	A	A	B	B	?
SPI	C	A	C	C	C	A	A	A	A	B	A	C	A	A	B	B	C	?
XTEST	C	A	C	C	C	B	B	B	B	B	A	C	A	C	A	B	C	?

Current models (see explanations below).

User Serviceability

There is typically nothing the user can do to service test indicators other than replace the contact point when it's worn. In some cases, the crystal can be replaced when it's discolored or scratched. Numbers refer to the scorecard above.

(1) **Crystal replacement**: (see page 124 for instructions) Only the new Mitutoyo and some Starrett indicators have plastic bezel and crystal combinations which simply snap in place. You may also need to replace the rubber o-ring.

- A = easy
- B = possible
- C = not possible without specialized tools

(2) **Contact point replacement**: (see information on contact point length on page 100) Only Starrett Last Word and Gem indicators may cause problems. Girod-Tast, Kafer and Starrett (other than Last Word) have a set screw which may have to be adjusted for calibration. All other other indicators are hassle-free. Manufacturers may provide special wrenches or keys which can make removal easier, but you can just as well use a pair of jeweler's pliers.

- A = easy
- B = might be tricky

(3) **Spare Parts availability**: Most manufacturers do not sell directly to end users. It may be necessary to order through a distributor. Distributors may not be eager to sell parts because they tend to be non-lucrative and time consuming.

- A = easily available through Long Island Indicator Service
- B = available from the manufacturer or distributor
- C = availability varies or is not available

(4) **Repairability**: Not only do parts have to be available, but the instrument has to be designed for possible repairs. Furthermore, technicians need to be skilled. While Long Island Indicator and many other repair shops can repair all of these gages, some are more suited to repairs than others.

- A = easily repaired by qualified repair shops
- B = requires manufacturer servicing
- C = not suited to repair

TECHNICAL CONSIDERATIONS

(5) **Accuracy**: All brands offer at minimum an accuracy of ± one graduation for their first revolution. Anything which claims to be more accurate is meaningless. Indicators with multiple revolutions become less accurate with each successive revolution. Since test indicators are comparators this again is meaningless. In-

cremental errors can be corrected to some degree by adjusting the contact point angle.

- A = excellent
- B = good
- C = fair

(6) **Repeatability**: This is a crucial aspect to test indicators used as comparators. Instruments which are dirty or damaged will show bigger variations in repeatability. This is a sign they should be sent in for servicing. Extended range indicators usually show insignificant variations. A new or recently repaired indicator will have the following repeatability:

- A = excellent
- B = good
- C = fair

(7) **Response**: Indicators with extended ranges have extra gears and you'll notice a minor sluggishness when compared to single revolution indicators. In practical situations this is insignificant. Interapid models 312B-15 and 312B-15V with very long points are often slower to respond because of the mechanics of the long point.

- A = excellent
- B = good
- C = fair

(8) **Magnetism**: Spinning motors and magnetic fields will magnetize indicators with iron content. This may cause the indicator to stick or freeze-up. Running the indicator through a de-magnetizer will be necessary. Most indicators are iron-free for this reason.

- A = non-magnetic
- B = will magnetize slightly and rarely inhibits function
- C = will magnetize to the point of non-function

(9) **Contact angle**: All test indicators except Interapid are designed so that the contact point must be parallel (180°) to the measuring surface. Deviations from this will result in a cosine error which can be mathematically compensated. Interapid indicators are accurate when the angle is 12°. Starrett model 708 and 709

must be 15°. These indicators are suited to situations where the body would otherwise get in the way. Some Fowler test indicators have pear-shaped contact balls which in theory eliminates cosine error.

- A = contact angle must be 180 degrees (parallel)
- B = contact angle must be some other angle (from 12° to 15° in some cases)
- C = contact angle is unaffected by cosine error (Fowler)

FEATURES

(10) **Dovetails**: test indicators need to be fastened to stands or holders of one sort or another. Most models have dovetails which are integral with the indicator body. This usually means they're of the same soft brass and they'll eventually disfigure from repeated clamping and tightening. Some models have dovetails made of hardened metal which is screwed onto the body. It would appear that these are replaceable, but you won't ever have to replace them for wear.

- A = replaceable dovetails (hardened)
- B = integral dovetails
- C = no dovetails

(11) **Plastics**: in order to save money some manufacturers are resorting to plastic parts, including plastic gears (Mitutoyo, Peacock) and plastic bezels. We find that the all-metal indicators are more durable. Plastic bezels are quick to break. Some structural parts may be made of fiber which is quite durable and is not considered as plastic.

- A = all metal (and/or with structural fiber parts)
- B = plastic bezel
- C = plastic gears or other parts

(12) **Craftsmanship**: some indicator construction shows real skill in manufacturing while others are downright amateurish. There can be variations among models of the same manufacturer. Good craftsmanship does not imply excellence for the task at hand. This is more a question of aesthetics.

- A = outstanding
- B = good
- C = shoddy

(13) **Manual vs. automatic reversal**: All modern test indicators can be used in either direction (contact point goes up or down). You can take measurements from below or from above. In the old indicators (ca. 1950-1960) there was a small lever on the side of the body which you had to move one direction or the other so that the indicator point moved up or down. This switch still exists on the Starrett Last Word and some other models. Problems will arise if the switch is not fully engaged. This can easily be overlooked and is a decided disadvantage.

- A = automatic contact point reversal
- B = manual switch on some models
- C = manual switch on all models

(14) **Body finish**: Indicators with a painted body can be problematic when the paint starts to peel, usually due to exposure to solvents. We see this occurring in many older indicators that come for repair. Interapid has apparently switched to a better paint job which doesn't peel like the older models did. Unfinished bodies tend to get rusty. The best indicator bodies have dull chrome plating on brass.

- A = plated body
- B = unfinished body
- C = painted body

(15) **Bezel rotation**: Manufacturers have tried different approaches to making the bezel movable and to keep it attached to the indicator body. The newest trend is to use o-rings which have the advantage that, generally, the bezel is easy to pry off and replace. But, o-rings, while they work well when new, age and stretch and the bezel won't stay put or turns too easily and can't be adjusted. The better systems use metal springs because they don't change appreciably. They can also be adjusted (although with difficulty) to customize the amount of friction. Often times these springs are hidden from view and have nothing to do with holding the bezel onto the indicator itself. This may very well be the best solution, as in the Interapid and Bestest indicators.

- A = independent metal spring (most reliable)
- B = metal spring which also holds the bezel
- C = rubber o-ring (least reliable, but easy to replace)

(16) Calibration: A few test indicators can be adjusted for calibration. Some have contact points with set screws so that, in effect, you can shorten or lengthen the point to adjust calibration. Others have internal cams that can be adjusted. Both procedures require jeweler's screw drivers, good eyes, patience and some intuitive algebraic skills. Since new indicators are factory calibrated, this feature is rarely needed by the end-user but could prove helpful in the repair shop.

- A = mechanism for adjusting calibration
- B = no means of adjusting calibration

(17) Waterproof?: Because the contact lever has to go into the indicator body, the test indicator can not be liquid proof. However, some models offer a bit more resistance to the ingress of liquids than others because the entrance into the body is smaller thus fending off liquid sprays; and, the bezels offer a tighter seal. If liquid contamination is an issue, then you may want to consider these factors.

- A = reliably waterproof
- B = a little bit better than nothing at all
- C = not waterproof at all

(18) Serial Number: To comply with ISO stipulations and to make identification possible for the employer's records of calibration a permanent serial number is required to be stamped or engraved somewhere on the indicator. Mitutoyo has gone so far as to add a miniature scan code on the dial face which can eliminate errors of transcription. If the dial is changed during repairs, however, the serial number is lost.

- A = permanent serial number on indicator body
- B = serial number and scan code on dial face
- C = no manufacturer's serial number

"Never memorize something that you can look up."
— Albert Einstein

BROWN & SHARPE BESTEST

This Brown & Sharpe indicator is still one of the best test indicators available. It is compact and has excellent repeatability. The black dial and orange hands have practically become a trademark for this gage. It's hard to imagine a better measuring instrument.

- black dial models: -5
- white dial models: -3
- yellow metric models: -13

There are several new improvements to the 2017 models. The indicator model number is now etched on the cover of the case rather than on the dial. Unfortunately because many of the components are sourced from foreign countries, the indicator can no longer legitimately be labeled "Swiss Made".

Swivel clamp 599-7045-1 is included with the horizontal models but also available separately. The clamp attaches to the indicator's dovetails. The clamp has two openings: 3/8" and 7/32" diameters.

Rectangular bar 599-7047 is included but also available separately. It is 1/4" by 1/2" by 3" long. The fastening stud is 7/32" (.220") in diameter and will clamp into the Swivel clamp shown above. The bar can then be fastened on your lathe or to your height gage.

BLACK DIAL BESTEST INDICATOR

Why are the black dials less expensive than white dials? Because they come with fewer accessories. You'll get a 1/4" diameter stem holder, contact point "wrench" (carbide .080" diameter point included) and a plastic box. Black dial models only come in the standard horizontal version.

BesTest 7033-5 is particularly popular. Notice that the graduations are sub-divided into .00005" but the accuracy is no better than the .0001" model.

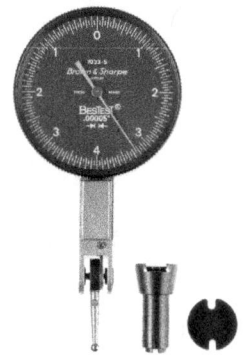

- The accuracy of .00005" and .0001" indicators is ±.0001"
- The accuracy of .0005" and .001" indicators is ±.0005"
- The repeatability of .00005" and .0001" indicators is ±.00002"
- The repeatability of .0005 and .001" indicators is ±.0001"

These test indicators come with an unsigned statement from the manufacturer that the gage is within tolerance and that these are traceable to "national master standards". There are no data provided.

Graduations	Range	Dial Ø	Point Length	Model
.00005"	.008"	1.5"	.5"	7033-5
.0001"	.008"	1"	.5"	7032-5
.0005"	.02"	1"	1.4"	7034-5
.0005"	.02"	1.5"	1.4"	7035-5
.0005"	.03"	1"	.5"	7030-5
.0005"	.03"	1.5"	.5"	7031-5
.001"	.03"	1"	.5"	7029-5

Horizontal BesTest Indicator

These are the standard, **white dial** variety (yellow dials if they are metric) with carbide contact points.

Bestest indicators have a very lightweight body and excellent response. They come with one carbide point (installed), with 2 accessories (shown below) and a plastic storage box.

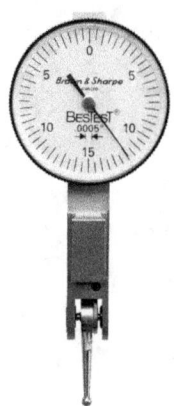

Note that the .00005" model dials are subdivided. Their accuracy remains at .0001"

- The accuracy of .00005" and .0001" indicators is ±.0001"

- The accuracy of .0005" and .001" indicators is ±.0005"

- The repeatability of .00005" and .0001" indicators is ±.00002"

- The repeatability of .0005 and .001" indicators is ±.0001"

Graduations	Range	Dial Ø	Point Length	Model
.00005"	.008"	1.5"	.5"	7033-3
.0001"	.008"	1"	.5"	7032-3
.0001"	.008"	1.5"	.5"	7023-3
.0005"	.02"	1"	.5"	7030-3
.0005"	.02"	1"	1.4"	7034-3
.0005"	.02"	1.5"	1.4"	7035-3
.0005"	.03"	1.5"	.5"	7031-3
.001"	.03"	1"	.5"	7029-3

VERTICAL BESTEST INDICATOR

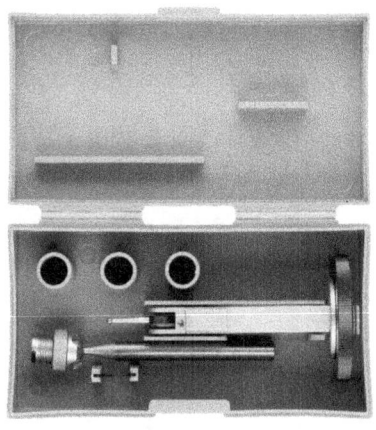

These are the standard, white-dial variety with the dial mounted on the end. This configuration is often referred to as **Jig-Bore Style**. Bestest indicators have a very light body and excellent response. These models come with a universal dovetail mounting attachment (see below), .080" diameter carbide point, contact point wrench, certificate of accuracy and traceability and an ugly plastic box. Other stem attachments, turning holders, accessories, contact points and spare parts are also available.

- The accuracy of .00005" and .0001" indicators is ±.0001"

BesTest does not make a metric vertical model but you can find an equivalent Tesatast indicator.

Universal holder 599-7054 is included with the vertical (end mounted dial) indicator but can also be purchased separately.

Graduations	Range	Dial Ø	Point Length	Model
.00005"	.008"	1.5"	.5"	7038-3
.0001"	.008"	1.5"	.5"	7024-3

BesTest Side Mounted Dials - Parallel Style

Ideally suited for work with height gages and transfer gages. The come supplies with a rectangular bar and swivel clamp.

Supplied in a plastic case together with:

- 1 contact point 1/2 in with a .080 in dia.
- 1 rectangular bar
- 1 swivel clamp
- 1 contact point wrench
- For metric versions, refer to Tesatast

Caution: if you are ordering from Amazon, check the description carefully since most of the time their photos are wrong.

Graduations	Range	Dial Ø	Point Length	Model
.0005"	.030"	1.5"	.5"	7021-3
.00005"	.008"	1.5"	.5"	7022-3

"It would be possible to describe everything scientifically, but it would make no sense; it would be without meaning, as if you described a Beethoven symphony as a variation of wave pressure." — Albert Einstein

Q: Why is the hand not set at 12 o'clock or 6 o'clock?

→ At rest, the hand is set at 5 o'clock on the BesTest indicator. This insures that you get the full range of travel. You will be adjusting the position of the indicator in such a way that the hand points to zero when you take readings. Now you will be able to read deviations in either direction. In the following example, you will be able to read .004" clockwise and .004" counter-clockwise, plus a little bit of over-travel.

Q: I am looking for a suction tool to remove crystals. I have several old B&S gages that I will need to replace the crystals; however, I am lacking the suction tool. Can you please let me know what it is called and where I might find one?

→ If you have really old models you won't need a suction cup but if you have the new ones we find that this method really doesn't work. You are better off breaking the crystal to install a new one. Detailed repair instructions are available in our Repair Manual for Swiss-made BesTest and TesaTast Indicators.

Q: I purchased several B&S 599-7032-5 about 3 months ago. Some of them are already sticking. I just think they are not jeweled and wear faster (this may be a total fantasy – but as you are the expert here, I thought I'd throw it out). Is B&S Bestest better than I think?

→ Sticking on the Bestest is due to only one of two things: oil has gotten inside and is causing the hair spring to malfunction or severe shock can break a tooth on the crown gear and cause the hair spring to coil up permanently. The Bestest indicator is perhaps a little more sensitive than the others but it also has excellent response and repeatability. It is fully jeweled. You can take a look by removing the cover plate (if you have a screw driver small enough). Unfortunately there is nothing you can do to clean it, except send it to a repair shop.

Q: I have a Brown & Sharpe BesTest 7031-5, and needs to have the crystal face replaced (cracked and cloudy). If I buy the replacement part, how do I change it? Or, do I need to send it out for repair?

→ You can replace the crystal if two conditions exist: 1. You are able to remove the old crystal. 2. The bezel is still in perfectly round, undamaged condition. In that case, the new crystal can usually be pressed in place with the thumbs.

Q: I believe that I have gotten in over my head so to speak. I have repaired these indicators in the past with good results, but have never replaced the hair spring before. Is it possible to purchase the bezel plate with the hair spring and anchor pin as an assembly?

→ This assembly is not available but you can find detailed instructions on hair spring repair, with photographs, in our Bestest repair manual.

Q: One of my coworkers bought a BesTest indicator two months ago and it is already malfunctioning. My best guess is at some point oil or moisture got inside and is making the needle stick and not repeat accurately. Is this covered under warranty?

→ Liquids are any test indicator's worst enemy, regardless of the brand, BesTest or otherwise. Once it gets inside, any indicator will stop repeating. If you can not keep liquids away from the indicator, then you should find another means of taking measurements. This can not be considered a warranty issue.

Q: I read that you guys recommend the white face on the Brown and Sharpe indicators unless people have experience with the black face. Why is this?

→ The red hand on the black dial may be hard to see for people with diminished vision or people who are color blind. Also, red is a hard color to see in dimmer lighting. Some red hands will wash out over time and become pale orange or even white. Sticking to the black hand / white dial avoids potential problems.

Q: I have a Federal Testmaster LT-111 which looks like it's the same as your 7033-5. Can it be repaired?

→ It is the same as 7033-5 and 7033-3 because it was made by the same manufacturer. It will use all the same parts and contact point.

BESTEST TEST INDICATOR CONTACT POINTS M1.4

Carbide, Ruby, Nylon and Teflon contact points for the Swiss made **Bestest** and **Tesatast** test indicators

The model suffix (-3 and -5) refers only to whether the dial is white or black. For example 7030-3 and 7030-5. Older, obsolete models had -2 or -1 as a suffix. They also use the same contact points. Points are identical for inch and metric models. Thread size is M1.4 (outside diameter is about 1.4 mm).

You can also buy a special double length point which is 1" long (28 mm from shoulder to center of ball). When installed, the readings on the dial will be doubled. In other words, the .0005" indicator will now read .001". Make sure you label your indicator when you do this, so that no one misreads the dial.

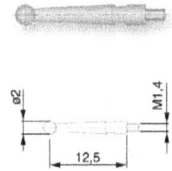

Tesatast uses the same points. Just check to see if they are the .5" or the 1.436" length.

How are they measured? From the shoulder (not including the thread) to the center of the ball. Custom ball diameters are available as well as teflon and nylon. Bestest contact points can be used with the Adapt-A-Tip adapter for dial and digital indicators.

Why are there multiple ordering numbers? The points are the same but the Brown & Sharpe (B&S) numbers are used in the USA, while the TESA numbers are for the rest of the world, while MTC is a numbering system we use. You may also encounter variations on the TESA numbers, such as 018602-00, 18.60202 or 01860202. (Notice that this merely eliminates the decimal point and adds a leading zero.)

CARBIDE, RUBY, NYLON AND TEFLON POINTS

These contact points are used for special applications, typically in optics.

- The **carbide point** comes standard with your new BesTest or Tesatast indicator. It is the most durable for general measurement.

- The **ruby point** outlasts standard chrome points and does not expand or contract with heat and cold. It does not conduct an electric current so it can be used on your machines without creating interference. It is also less likely to scratch delicate surfaces. It's also the material that will hold up to silicon carbide surfaces.

- The **Teflon point** is softest made of a solid Teflon ball. It is very white in appearance. It is least likely to scratch but is not as long lasting as the others. The smallest Teflon diameter available is 1/16" (.062"). Smaller points can not be manufactured in this material.

- The **Nylon point** is not as soft as Teflon but will last longer. It can also be used where surface damage is an issue. Nylon appears opaque.

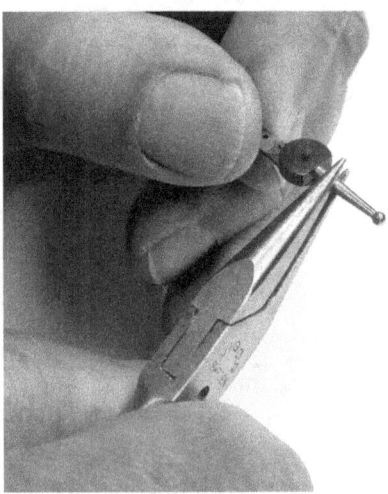

Use small pliers to tighten the contact point, or use the little wrench that came with your indicator.

M1.4 Standard Points (.5")

Designed to be used with the following models:

- 7021-3
- 7022-3
- 7023-3
- 7024-3
- 7029-3
- 7029-5

- 7030-3
- 7030-5
- 7031-3
- 7031-5
- 7032-3
- 7032-5

- 7033-3
- 7033-5
- 7033-13
- 7037-3
- 7038-3

Ball material	Ball Ø	Length	B&S equivalent	TESA equivalent	Our Part Number
carbide	.015"	.5"	599-7030-15	18.60200	MTC21-07
carbide	.040"	.5"	599-7030-08	18.60201	MTC21-08
carbide	.080"	.5"	599-7030-80	18.60202	MTC21-10
carbide	.120"	.5"	599-7030-120	18.60203	MTC21-12
ruby	.040"	.5"	—	18.60205	MTC21-18
ruby	.080"	.5"	—	18.60206	MTC21-19
Teflon	.060"	.5"	—	—	MTC21-22
Teflon	.080"	.5"	—	—	MTC21-23
nylon	.060"	.5"	—	—	MTC21-25

Contact points can be bought online by visiting our website:
www.longislandindicator.com

M1.4 Double Length Points

You can also buy a special double length point which is 1.1" long (28 mm from shoulder to center of ball). When installed, the readings on the dial will be doubled. In other words, the .0005" indicator will now read .001".

Make sure you label your indicator when you do this, so that no one misreads the dial.

For the following models only:

- 7021-3
- 7022-3
- 7023-3
- 7024-3
- 7029-3
- 7029-5

- 7030-3
- 7030-5
- 7031-3
- 7031-5
- 7032-3
- 7032-5

- 7033-3
- 7033-5
- 7033-13
- 7037-3
- 7038-3

Ball material	Ball Ø	Length	B&S equivalent	TESA equivalent	Our Part Number
carbide	.030"	28 mm	None	None	MTC21-34
carbide	.080"	28 mm	None	None	MTC21-37

M1.4 Long Points (1.436")

Designed to be used with the following models only:

- 7034-3
- 7034-5
- 7035-3
- 7035-5

Ball material	Ball Ø	Length	B&S equivalent	TESA equivalent	Our Part Number
carbide	.015"	1.436"	—	18.60210	MTC22-02
carbide	.040"	1.436"	599-7034-40	18.60211	MTC22-04
carbide	.060"	1.436"	—	18.60216	MTC22-05
carbide	.080"	1.436"	599-7034-80	18.60212	MTC22-06
carbide	.120"	1.436"	599-7034-120	18.60213	MTC22-08
ruby	.040"	1.436"	—	—	MTC22-15
ruby	.080"	1.436"	—	—	MTC22-16
ruby	.100"	1.436"	—	—	MTC22-17
Teflon	.060"	1.436"	—	—	MTC22-19

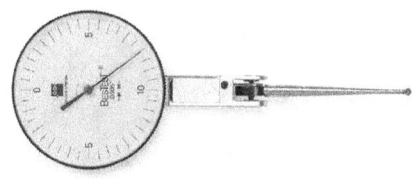

New indicators no longer have the model number on the dial. Look for it on the side of the body instead.

TESATAST AND SWISSTAST

Points are identical for inch and metric models. For that matter, these points are identical to Bestest points. You can order any of the Bestest points but be sure to order the correct length as shown below.

Tesa model	Contact point length
18.10005 18.10006 18.10009 18.10010 18.10013 18.10204 18.10304 18.11001 18.20006 18.20007 18.20010 18.20011 18.20012 18.20013 18.20014 18.20015 18.20204 18.20304	.5" (12.5 mm)
18.10007 18.10008 18.10012 18.10205 18.20008 18.20009 18.20016 18.20205	1.436" (37 mm)

FEDERAL

The discontinued Federal Series 100 indicators (T-101, T-102, etc.) were manufactured by TESA in Switzerland. The contact points are identical to those used by the Brown & Sharpe Bestest indicators.

Old style points (M-1, T-1, etc.) had ratchets rather than threads. They are no longer available from us.

GIROD-TAST: THE OTHER SWISS TEST INDICATOR

Girod-Tast indicators are manufactured in Court, Switzerland by BG Instruments SA, a family-run business founded in 1987, with approximately 40 employees. Whereas the look-alike Bes-Test and Tesatast indicators are mostly made in China, the Girod-Tast can still claim to be "Swiss Made".

Girod, pronounced "she-row" (Girod-Tast, Girod-Tast) is distributed by **Fred Fowler** in the USA and offered in a number of big-house catalogs. In Switzerland these are also available under the name of SISO-tast. It's a good indicator, made in Switzerland, structurally similar to the old model TESA indicators. They have also introduced long range models with multiple revolutions and an improved sturdier crown gear.

Although inch reading versions are available, these indicators would be an ideal choice if you are working in a **metric** environment. And, as per AGD standards, the metric dial will be yellow if you bought the gage from Fowler. (In the rest of the world, the metric dial will be white.)

Girod-Tast indicator specifications

Indicator type:	DIN	.0005"	.0001"	0,01 mm	0,002 mm
Deviation:	fe	.0005"	.0001"	10 µm	2 µm
Total deviation:	fges	.00065"	.00016"	13 µm	3,5 µm
Measuring inversion:	fu	.00015"	.00006"	3 µm	1,5 µm
Repeatability:	fw	.00015"	.00004"	3 µm	1 µm

A NOTE ABOUT THE GIROD-TAST CONTACT POINTS

Here is a simple way to tell whether you have the old or new model Girod-Tast.

Your old models will have the same black bearings on both side of the body. Your new models will have different bearings on either side.

Old models (pre-2008) contact points are held securely in place with a set screw which is inserted from the back end. Swivel the point to the side and you'll see it. A very small screwdriver is needed to loosen this set screw by a half turn. This allows you to unscrew the contact point.

Old model contact points had M1.6 threads and these will not fit the new GL series.

Many Girod contact points have a strange burr above the thread. You may find that you can't insert the new point without stoning off this burr. When the new point is screwed in, you'll want to tighten the set screw again. Notice that, if the set screw is too far out, you may not be able to swivel the point back into position.

This same set screw is used to shorten or lengthen the contact point so that the indicator calibrates correctly. It's quite a nuisance. You'll have to calibrate the indicator and if the readings are over by one graduation or more, you'll want to lengthen the point. Do this by screwing it out, by 1/2 turn at a time and then tightening the set screw. Calibrate again, and again, and again until you get it right.

Why did the manufacturer torment us with this "feature"? If the contact point is used at an angle you will have to make mathematical adjustments for the cosine error. However, on this kind of indicator you can permanently adjust the length of the point so that readings at a particular angle will always be correct.

New models (since 2008) of the "series GT" have eliminated this setscrew making adjustments and mathematical calculations unnecessary. They now operate on the same principle as the Bestest and Tesatast indicators. New model contact points have M1.4 threads and will not fit the older models.

How can you tell whether it is M1.4 or M1.6? Just take your old point and measure the outside diameter of the threaded portion metrically. M1.4 threads will measure 1.4 mm in diameter.

GIROD-TAST INDICATOR REPAIR

Spare parts **should** be available from Fred Fowler Company but are sometimes in short supply and not all repair shops will be able to provide repair service. It's a little bit like buying a car in a part of your state where the nearest service center is 120 miles away. If future service is anticipated you may want to ask your distributor for available options.

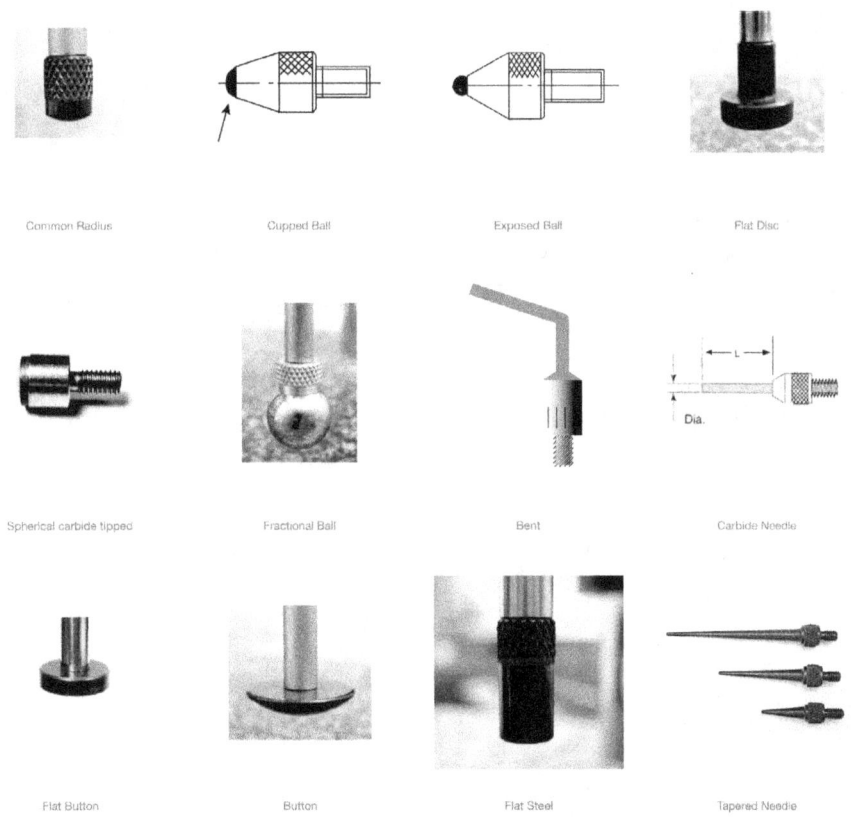

A complete selection of Dial Indicator contact points is also available on our web site.

COMPAC SWISS TEST INDICATOR

Sadly, all Compac test indicators were discontinued in 2020. Since we still repair indicators that were bought in the 1960s there is every reason to believe they will remain in the connoisseur's tool box for years to come.

Compac models are designed to permit a true reading when the angle between the stylus and the reference surface is 0° (i.e.., they are parallel).

Since the dial test indicator is used as a comparator, the angle can usually be set to remain at 0° for all comparisons. However, when the indicator is run through its range while attempting direct readings or for calibration, this angle will of necessity change, introducing the cosine error.

In the rare cases when measurement is required with the stylus at an angle other than 0° with respect to the reference surface the readings should be corrected. Refer to calibration instructions for details.

All new Compac test indicators come with a calibration certificate from the manufacturer, but they are not NIST traceable. If you need an NIST calibration certificate, please specify at time of ordering. We will provide these certificates for an additional charge (series 210, 220 and 230 indicators shown on this page only).

While Compac test indicators are accurate ± 1 graduation when used as a comparator, they have total allowable deviations over their entire range as follows:

Compac model	total range	dial configuration	total deviation	hysteresis	repeatability	measuring force
214A / 214GA	.060"	0-10-20	.0005"	.00015"	.00015"	0.35N
224A / 224GA	.060"	0-10-20	.0005"	.00015"	.00015"	0.35N
234A / 234GA	.060"	0-10-20	.0005"	.00015"	.00015"	0.35N
213LA / 213GLA	.120"	0-20-40	.001"	.00025"	.00015"	0.20N
223LA / 223GLA	.120"	0-20-40	.001"	.00025"	.00015"	0.20N
233LA / 233GLA	.120"	0-20-40	.001"	.00025"	.00015"	0.20N
215A / 215GA	.024"	0-20-40	.0005"	.0001"	.00005"	0.30N
225A / 225GA	.024"	0-20-40	.0005"	.0001"	.00005"	0.30N
235A / 235GA	.024"	0-20-40	.0005"	.0001"	.0001"	0.30N

Source: Tesa SA, July 2005

Q: We use a test indicator to qualify radial runout of the work rolls on a cold rolling mill. This check is made with the machine in 'jog' – so that the work rolls are preloaded. Unfortunately, the test indicator is exposed to the oil used for coolant. The one we are using now has oil inside the dial, and we are looking to replace it.

➡ As you noticed, test indicators are not coolant proof, and many indicators come here for repair because the hair springs have become oily and stop functioning. Your indicators can be sent for cleaning as well. The cleaning bill will be less for single revolution indicators so you might want to buy only those models in the future. Bestest or single revolution Compac indicators may be a good choice. You could also opt for a disposable indicator made in (gasp!) China. You can probably buy them for less than the cost of repair and just throw them away when they stop functioning. Repairs (even cleaning) will not be possible on those.

Q: Do you have or know of any needles for an Interapid test Indicator that are not attracted to a magnet? I know there are ones where the tip is ruby or carbide, but I am looking for something where the whole needle is immune to magnetism.

➡ A: Titanium is considered paramagnetic which means it holds negligible amounts of magnetic charges for very brief intervals. We don't have any of these for the Interapid indicator but they would work on Mitutoyo models 513-212, 513-412 and 513-446.

"I suppose it is tempting, if the only tool you have is a hammer, to treat everything as if it were a nail." — Abraham Maslow, TOWARD A PSYCHOLOGY OF BEING

INTERAPID TEST INDICATOR

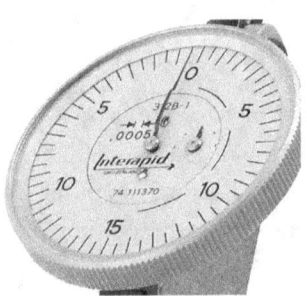

This genuine, Swiss-made Interapid indicator is undisputedly the best test indicator of its kind. The large hand makes two complete revolutions in either direction and the small hand helps you keep track which revolution you're on. The right hand side of the dial has a thin black half-circle in case you want to use these in a set-up involving mirrors.

The 4 mm diameter mounting stem swivels and can be replaced if damaged. The two dove tails (back and top) are mounted with screws and these can also be replaced or removed if they get in the way. These dovetails, by the way, are hardened metal and won't get squashed with wear like the brass dovetails on other indicators. If you don't want 4 mm, then you can easily slip on a 3/8" adapter.

Many different contact points, including ruby, Teflon and nylon are optionally available, and easy to install. The contact point operates on a friction clutch, so you'll have to apply a bit of force to get the point to swivel, which it does, from the front of the indicator to the back. New indicators come with a special "wrench" for removing the contact points. It's okay to use a pair of jeweler's pliers for this purpose.

Unless you know what you're doing, you should under no circumstances attempt to disassemble or adjust any of the screws or bearings on the indicator.

These indicators are designed to function when the contact angle is about 12° to the work surface.

These indicators come in a sturdy box, with a .080" diameter carbide point installed, a wrench for removing the contact point and a manufacturer's statement of conformance along with allowable errors (see bottom of this page). Calibration certificates are not available. If you require such, you will have to send the indicator to an accredited calibration lab in your area.

Special 1.45" long contact point to be used in place of any .687" point. When you install this point, the readings on your dial will be doubled. That is, the .0005" graduations will become .001", the .001" will become .002", and the .0001" graduations will become .0002"

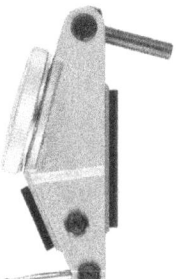

The contact point and the holding stem can both be repositioned as needed.

Graduations	Range	Dial Ø	Point Length	Horizontal Model No.	Order No.
.0005"	.06"	1.5"	.687"	312B-1	74.111370
.0005"	.06"	1.5"	2.750"	312B-15	74.111965
.0005"	.06"	1"	.687"	312B-2	74.111371
.001"	.06"	1"	.687"	312B-20	74.111374
.0001"	.016"	1.5"	.687"	312B-3	74.111372
.0001"	.016"	1"	.687"	312B-4	74.111373
0,01 mm	1,6 mm	37,5 mm	16,5 mm	312-1	74.111366
0,01 mm	1,6 mm	30 mm	16,5 mm	312-2	74.111367
0,002 mm	0,4 mm	37,5 mm	15,2 mm	312-3	74.111368
0,002 mm	0,4 mm	30 mm	15,2 mm	312-4	74.111369

Indicators can be bought online by visiting our website:
www.longislandindicator.com

Vertical Interapid Test Indicator

Vertical, jig-bore type test indicator with one dovetail and an integral 4 mm holding stem which swivels in one direction. You can buy adapters so that the diameter of this stem becomes 3/8" or 8 mm. (See below)

The contact point (Ø.080" supplied) also swivels through a wide arc. These are available from a short .687" to a very long 5.5" length. Other diameter balls and other lengths as well as contact points made of Nylon, Teflon or ruby are optionally available.

A friction clutch holds them in a steady position so it might take a bit of effort to move the point. Do not use any tools other than your fingers for this operation and be sure that the point is securely fastened before attempting to move it.

The vertical models have just one dovetail.

Graduations	Range	Dial Ø	Point Length	Vertical Model No.	Order No.
.0005"	.06"	1.5"	2.750"	312B-15v	74.111958
.0005"	.06"	1.5"	.687"	312B-1v	74.111377
.0005"	.06"	1"	.687"	312B-2v	74.111378
.001"	.06"	1"	.687"	312B-20v	74.111379
.0001"	.016"	1.5"	.687"	312B-3v	74.111957
0,01 mm	1,6 mm	37,5 mm	16,5 mm	312-1v	74.111375
0,01 mm	1,6 mm	30 mm	16,5 mm	312-2v	74.111376

Indicators can be bought online by visiting our website:
www.longislandindicator.com

COMPLETE INTERAPID INDICATOR SETS

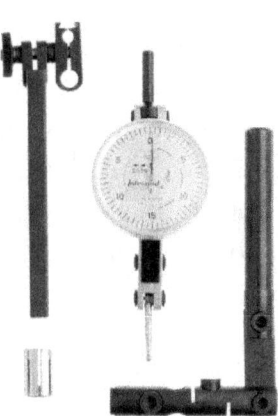

Several models are available in complete sets. They will also include the following:

- 74.106331 height gage bar with clamp
- 74.108943 adapter sleeve 3/8"
- 74.108942 adapter sleeve 8 mm
- 74.106931 axial support 3/8" Ø stem
- 74.106026 axial support 8 mm Ø stem

These parts are described individually on our indicator attachments page.

Interapid Model No.	Style		Complete Set Order No.
74.111370	312B-1	.0005" horizontal	74.111508
74.111371	312B-2	.0005" horizontal	74.111509
74.111372	312B-3	.0001" horizontal	74.111510
74.111377	312B-1v	.0005" vertical	74.111513

"We are stuck with technology when what we really want is just stuff that works." — Douglas Adams, THE SALMON OF DOUBT

IMPORTANT CALIBRATION INFORMATION

Unlike other test indicators, Interapid indicators do not have serial numbers on them. Perhaps someday the manufacturer will remedy this situation but for now you'll have to assign your own unique serial number to each indicator if you intend to keep track of their calibration. It's easy to scratch a number into the painted body using a sharp tool. A permanent marker with a super fine point will also work. Do not use an electro-engraver because it creates sparks which have been known to damage the pinions on the gears.

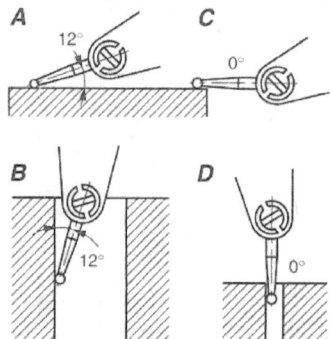

Interapid series 312 test indicators are designed to permit a true reading when the angle between the stylus and the reference surface is 12° as shown in illustration A and B.

In the rare cases when axial measurement is required with the stylus at 0° with respect to the reference surface as in C and D in the illustration, (for instance: small diameter bores) the readings must be multiplied by a factor of 1.022

It also follows that incremental calibration discrepancies can be corrected by adjusting the contact point angle.

Here are the most recent specs as published by the manufacturer (specifications for older models may be slightly different):

Model	Graduations	Deviation per revolution	Deviation Total	Repeatability	Hysteresis	Measuring force
74.111366 74.111367 74.111375 74.111376	0.01 mm	10 μm	23 μm	3 μm	3 μm	0,12 N
74.111368 74.111369	0.002 mm	4 μm	10 μm	1 μm	2 μm	0,25 N
74.111370 74.111371 74.111377 74.111378	.0005"	.0004"	.00092"	.00012"	.00012"	0,12 N
74.111372 74.111373 74.111957	.0001"	.00016"	.0004"	.00004"	.00008"	0,25 N
74.111374 74.111379	.001"	.0004"	.00092"	.00012"	.00012"	0,12 N
74.111958 74.111965	.0005"	.0006"	.0016"	.00004"	.0008"	0,12 N

Source: Tesa specifications (1899.024.1605) which are included with each indicator as of April, 2017.

INTERAPID INDICATOR FEATURES

These indicators have a very light measuring force: just 0,12 N for .0005" models and 0,25 N for .0001" models.

The indicator is designed to measure with the contact point held at an angle of 12° to the measuring surface. (Some sources list this angle as 13 degrees. The difference is not significant.) This allows you to take measurements down a hole or in a bore.

The .080" diameter carbide contact point is easily replaceable with other diameters (from .015" to .120") and other substances (ruby, teflon or nylon). Under some circumstances, you can even install a longer point. See contact points.

The dial hand makes two revolutions in each direction. The small counter hand keeps track of the revolutions.

The dial has balanced numbering: 0-4-0 (.0001" indicators) and 0-15-0 (.0005" indicators). The right half of the dial features a thin black band which will help identify the plus or minus side when using a mirror.

The novel tapered shape tilts the dial towards the observer. Both Mitutoyo and Starrett have recently copied this design. This makes for easier reading with less likelihood of parallax errors as well as a reduction in glare from overhead lights.

If you need to slip the indicator into an opening then you may be interested to know that the width of the body at the stylus (contact point) is 10,3 mm which is broader than Bestest, Tesatast or Compac.

The holding stem is 4 mm (5/32") in diameter. The stem swivels and is attached to the body, although it can be removed and replaced. Simple instructions for this procedure are available on our Do-It-Yourself page. If you need a holding stem with a different diameter, use one of the following adapters show on our spare parts list:

- An elegant 3/8" adapter fits over the 4 mm stem. Order no. 74.108943
- For metric set-ups, use the 8 mm stem adapter. Order no. 74.108942

The dovetails are hardened metal, screwed onto the body. These can also be removed and replaced. See our spare parts list for replacement parts.

INTERAPID REPAIR

Long Island Indicator is an authorized repair service for Interapid test indicators.

Top-of-the-line Interapid test indicators are probably the most challenging of all to repair. Absolute cleanliness and the replacement of critical parts is essential to a quality repair. We've been trained by masters in the craft, and having an ample supply of fresh parts directly from the manufacturer are able to return any Interapid test indicator to A-1 condition.

Is the indicator damaged beyond repair? If the body is damaged in any way (the front end is squashed, for instance) then we'd have to replace the body and you'd have to pay about $150 for the repair. Not usually worthwhile. If the whole face has come off because it was ripped off - that sounds serious, yet happens regularly - it's repairable for the normal cost quoted in the paragraph above.

Sorry, we can't repair old models such as the 310B and 311B series. Parts for these indicators are no longer available. If have one of these, then you've gotten your money's worth since they were discontinued about 60 years ago. If you just bought one on e-bay, and it doesn't work, then you've been had.

Is Interapid any good?

We rate them among the very best. If you'd like to compare the good points and the not-so-good points of the Interapid indicator with other manufacturers then take a look at our indicator comparisons page.

Using the correct contact point

While various diameter balls do not affect readings, the correct contact point length is critical. (Refer to model information on this page.) Use only points designed for Interapid indicators. Any deviation will result in reading errors.

When ordering new points, take note of the indicator's model number and decide on the ball diameter you'll need.

Problems with repeatability

According to the manufacturer, the Interapid test indicator is allowed a repeatability error as shown in the reference chart above.

Repeatability must be checked against the measuring surface. When the contact point is resting against the measuring surface, turn the bezel so that zero is directly in line with the pointer (hand). The indicator's "at rest" zero reading must not be used to verify repeatability.

Swiveling the contact point may cause the zero setting to change slightly. If this happens, reset zero by turning the bezel.

Zero readings may also be off when comparing contact point up and contact point down motions. Reset the zero by turning the bezel as needed.

Furthermore, repeatability can only be assured if the indicator is securely fastened to a good test stand or other fixture. Fastening by the dovetail as close as possible to the front of the indicator is the preferred method.

INTERAPID OLDER MODELS

Current test indicator models begin with the number 312. Previous models dating to the 1960s were number 311 and these are often still repairable, depending on the damage. The oldest models were number 310 and you can assume these are beyond repair simply because spare parts are obsolete. If the horse-shoe shaped clip which holds the bezel in place is missing, then you may have to improvise with a piece of spring-wire bent into the correct shape. You might be tempted to say "It only needs a cleaning," but that's almost never the case. None-the-less, if you want us to have a look, it costs you nothing but shipping.

For a brief period of time Brown & Sharpe had put its name on the Interapid indicators. For example, B&S model 7025-4 was actually Interapid model 312B-1. These can still be repaired because they were not imitations but genuine Interapid indicators made in Switzerland.

INTERAPID DOVE TAILS AND STEM ATTACHMENTS

It would appear that one of the advantages Interapid indicators have over other brands is the replaceable dovetails. Each of the 1 or more dovetails (depending on models) is fastened to the body with several screws. Dovetails are made of hardened steel and they won't break or wear down like those which are integral with the bodies of Bestest, Mitutoyo, Compac and others. Because they're hardened, they probably will never have to be replaced. Unless... the entire dovetail has been ripped from the body through some misadventure. In this case, the screw holes are usually damaged beyond repair anyway.

The feature of replaceability is not particularly important. However, the hardened steel is. These dovetails will outlast any others. If you'd like to buy replacement dovetails and screws check the Interapid parts list.

Do not be tempted to adjust any of the ball bearings. Their proper adjustment is crucial to the indicator's function. If you are attempting a do-it-yourself repair, please consider checking our **Interapid Repair Manual** first.

Q: Is there a way to check and adjust the Interapid "stem" to contact point concentricity? The machinist in our mill department often uses the Interapid to centralize the spindle to the centerline of a hole in the part by sweeping the hole.

➡ There is no way this can be adjusted nor was the indicator designed for this purpose.

Q: What difference is there between the 312B-2V which lists as a .0005" model and the 312B-20V which lists as a .001" model? Is there any difference other than the additional lines (between the .001 lines) printed on the dial of the 74.111378 model?

- ➡ There is no difference except the one that you suspect. The indicators are identical in every other way.

Q: I was reading on your website that there are counterfeit Interapid test indicators. Do you know how to tell the difference or do you have any information on identification of the fakes.

- ➡ Although they look very much alike, the Chinese indicators DO NOT have the name Interapid on them. They are sold with names such as X-Test. All genuine Interapid indicators have the Interapid name printed on the dial

Q: I have two Interapid indicators. Both of these indicators seem to not read as close as my Brown & Sharpe Bestest indicators. Both Interapid indicators are 99.85% good and clean but I would like them to be a little smoother and trustful in the reading of them.

- ➡ There are two differences between the Bestest and the Interapid indicators. You must keep these two points in mind when comparing them. They will never be equal in responsiveness nor in accuracy.

    1. The Interapid indicator is only accurate when the contact point is at 12-degrees to the work surface.

    2. The Interapid indicator has more moving parts meaning that it will be slower to respond and will have poorer repeatability, although these differences are almost negligible.

Q: My Interapid works fine but it has gotten very dirty. Don't tell me to wipe it with a paper towel because that won't work. Any suggestions?

- ➡ Dip a toothbrush in some Naphtha[1] and scrub away. Avoid getting solution inside the indicator. If Naphtha doesn't remove stains, use Simple Green or Fantastik on a paper towel (!), just be doubly sure you don't get any of these cleaners inside.

---

[1] My doctor was disturbed to hear that I've been working with Naphtha for over 40 years. The benzene component is considered cancer causing. You might want to consider an alternative or at least speak with your doctor about the possible consequences.

Q: You mention that the large hand of an indicator should be at 9 or 10 o'clock. What if it rests between 12 and 1? Is that a malfunction?

- On the Interapid, the hand is at high noon when at rest. If it is slightly off, but is still accurate, then there is nothing to fret about. But, if the hand drifts or returns to various positions, then it should be checked out.

Q: Have you have seen any increase in warranty claims and/or quality issues with Interapid indicators 312B-1, -3, & -20?

- Just the opposite, over the past 12 months we have seen a decrease since the manufacturer fixed a "bug" which had recently surfaced.

Q: If I purchase a new Interapid through an eBay seller, is the warranty honored by the manufacturer?

- If the item is new when purchased, then TESA will honor the warranty. You would have to keep the purchase receipt to prove the date of purchase. It does not matter if the seller is an authorized distributor or not.

Q: Contrary to my nature, I thought I'd ask for advice before I screwed up something. After all, you do say "Unless you know what you're doing, you should under no circumstances attempt to disassemble or adjust any of the screws or bearings on the [Interapid] indicator." However, on your "Do It Yourself" web web page you say this about Interapid stem attachment: "You will notice that you can increase the swivel friction by tightening screw A." Can the swivel friction on the contact point be adjusted in the same way? The contact point on mine doesn't want to swivel, and I'm worried that I shouldn't push any harder on it. Also, should I put any oil on this friction clutch?

- This is a good question and I'm glad you asked. The contact point friction can not be adjusted in any way. The Interapid has fairly hard friction on the point. After all, you don't want it to move on its own. Don't be afraid. The contact point angle can be adjusted. Just use your thumb and the point will yield. A tiny drop of oil can be put on the "clutch" if you still feel that there is too much resistance.

Q: My Interapid seems a little jumpy, especially when changing direction. Is it in need of repair?

> Jumpiness is most often the result of a dirty hairspring. When the spring gets oily, it stops functioning and needs to be cleaned. This is easier said than done and should be put in the hands of a professional.

Q: We use the 312B series Interapid here on a daily basis. A few of them from time to time get abused by inexperienced or negligent machine operators. My question to you is, is there a way to zero out the small dial (rev counter) on these things? Repeatability does not seem to be a problem, I just would like to keep the needles in line as they were from factory.

> This should really not be happening. I suppose, if the large hand remains in its 12 o'clock position but the small hand has shifted, then the small hand might be loose on the pinion. You can unscrew the bezel and, if the small hand is loose enough, simply align it into the correct position and tap it down (we use a staking tool). If you feel much resistance though, I suggest you don't play with it.

Q: The contact point moves up and down (forward and backward is a better description) but one direction is noticeably harder than the other. Shouldn't they be the same?

> Ideally—yes—and probably in a new indicator this will be the case. Interapid indicators have a spring mechanism that controls this force which is quite different from any other manufacturer's. Unfortunately it can not be modified without taking the entire indicator apart. Don't attempt this unless you already know how to put it back together! The spring can weaken in one direction over time but the problem is more likely the result of previous repairs. Just the act of taking the mechanism apart can distort the spring's shape. The difference in force shouldn't matter much but if it's imperative to your operation that they are identical, you should consider switching to a different brand. Bestest, Tesatast and Compac would be ideal substitutes.

Q: I know you have been asked before, but I really don't like the peeling paint on my Interapid. Isn't there something I can do?

> Of course there is. You will have to completely disassemble the indicator until you have just the body left. Use paint stripper if it still has bits of paint on it. Then spray with a primer followed by a color of your choice. Be careful you don't get paint onto any of the screw threads or

you'll have a heck of a time trying to get the screws back in. Here's a handy hint: if you don't mind spending the money, just buy yourself a new body instead. Then your indicator will look like an original and the paint on the new bodies will last much longer.

Q: I have a model 310-B1. I see that parts are no longer available. The only thing wrong with the indicator is that the face comes off easily. The plastic split-ring that holds it on needs to be replaced and I was hoping that there might be another indicator that incorporates the same ring that I could purchase. Please let me know if that is the case.

- Those are exactly the parts that are the weak link on this indicator and those rings are long gone... no one else uses anything like it. The other missing item is the little horse- shoe shaped clip. If you really love this indicator, We suggest you try to improvise a way to hold the bezel in place. A properly shaped piece of spring wire might do the trick.

Q: We recently purchased a new Interapid 312B-3 indicator and noticed that the dial appears to slightly lift from the base. Is this common with these indicators ?

- The movements on the large diameter indicators are mounted on a spring washer which creates the friction to keep the dial from rotating unnecessarily. There is some wiggle as a result. You should notice it when you "pull" the bezel upwards. It should not be more than about 1 mm.

Q: I would like to know if the 2.75" #111913 indicator point can be used with the 312B-3 Interapid indicator? I believe that the divisions will now be 0.0005" instead of 0.0001" with this longer point.

- The readings would be very close to .0004" per graduation but the indicator will most likely stick, jump and hang with this size point installed. The hair spring can not handle the extra weight of this long point. It is not advisable.

Q: Can I get the small wrenches to change the Interapid contact point properly? The tips all have the small flat machined into the shank and a new indicator used to have one of these wrenches supplied with the kit but I haven't seen one with any of the new indicators we've purchased recently. It used to be a simple round disk with a notch that fit the flat on the tip shank. Without this a needle-nose pliers works but seems wrong putting a pliers on something so delicate.

→ I'm surprised to hear that these wrenches are missing in some of your new gages. They really should have been included but they do tend to fall out. Someone may have opened the boxes at some point (even during shipping) and may not have known the significance of this little wrench. By the way, we use small pliers. They work very well. (The wrench is part number 01860307 and can be ordered on web page 151.)

Q: I purchased a spring and gear assembly for the Interapid indicator (312B-3) and I am having a difficult time getting the indicator to return to 0 accurately (within .0001). Part of the problem is adjusting the ball bearing with spanner nuts. Will you please help me with this?

→ Interapid indicators are difficult to repair. It takes most of us several months of practice to get it right. I suggest you send it to a repair shop for servicing.

Q: I have a model 310-B1. I see that parts are no longer available. The only thing wrong with the indicator is that the face comes off easily. The plastic split-ring that holds it on needs to be replaced and I was hoping that there might be another indicator that incorporates the same ring that I could purchase. Please let me know if that is the case.

→ Those are exactly the parts that are the weak link on this indicator and those rings are long gone... no one else uses anything like it. The other missing item is the little horse-shoe shaped clip. If you really love this indicator, I suggest you try to improvise a way to hold the bezel in place. A properly shaped piece of spring wire might do the trick.

Q: I have an Interapid model 312-b3 indicator. I need to use it to center a rifle barrel (center the interior bored hole) in a lathe at a point fairly deep into the hole. I believe I could use your long indicator points for this operation as long as I'm aware that the accuracy of the dial is decreased.

→ If the contact point is exactly twice the length of the standard point, then the graduations on the .0005" indicator will equal .001" and the graduations on the .0001" indicator will equal .0002".

Q: I recently purchased an Interapid 312B-1V (used, no directions). In order to move the contact you state: "be sure that the point is securely fastened before attempting to move it." Would this mean to ensure point is tight in holder then run the contact against the stop? Then move to desired angle?

- As you've probably noticed, the contact will swivel (with a bit of force applied) in two directions. Just make sure that you've completely screwed in the contact point before you swivel otherwise you might break off the point in the process.

Q: I have an older Interapid test indicator, model # 311B-1 and I was wondering if by any chance you would still have parts for it? The part I'm looking for is the one that holds the bezel to the extension ring. It looks like a horse shoe, or if you know of any company that makes replicas such as this.

- We called it the horseshoe clip at the time. No one makes this part as far as I know. People have been known to improvise with a bit of copper wire which can pretty easily be bent into shape and filed down if necessary.

Q: I have a brand new Horizontal model 312B-15 with 2.75" long contact point. I ordered the wrong one, so I bought some shorter tips for this indicator, now it acts jerking and does not repeat with a shorter tip on it. Does this mean that I can only use the long tip?

- The long point models have a different hair spring to compensate for the weight of the point. Apparently replacing this with a short point causes the jerkiness you've noticed. The only thing you can do is stick with the long point, or have the hair spring replaced.

Q: The paint is peeling from my Interapid indicator. Can I do anything?

- This problem is seen more commonly among the older models. Certain solvents softened the paint and it eventually began to peel. We don't see the problem anymore on newer models. My suggestion is to scrape off the peeling paint. You'll end up with a shiny brass body which isn't so bad looking after all. Trying to repaint is usually an invitation to failure. For one thing, you'd have to disassemble the indicator completely and unless you really know what you're doing... please don't!

Q: How could I sterilize an Interapid indicator used in biological experiments?

- Surface cleaning with isopropyl alcohol can be used, just don't immerse. There is a small amount of oil in the bearings and we have to be sure that these don't dry out. Dried oil becomes gummy and that would be detrimental. For complete sterilization, Gamma radiation may work very well. I don't see how it could damage any of the gage's components as long as no excessive heat is produced. Having never tried this, though, it is only a guess. Ethylene oxide can be used if it doesn't dry the oil. I don't know if it would. If the crystal becomes fogged by the gas, you'd have to replace it, but that would be part of the cost of sterilization. Autoclaving to 120-degrees is probably too severe a method. The presence of steam would be quite damaging. I would avoid this method unless dry heat can be used. In summary, use the alcohol whenever possible, before resorting to any of the other methods.

Q: The contact point moves up and down (forward and backward is a better description) but one direction is noticeably harder than the other. Shouldn't they be the same?

⇒ Ideally yes and probably in a new indicator this will be the case. Interapid indicators have a spring mechanism that controls this force which is quite different from any other manufacturer's. Unfortunately it can not be modified without taking the entire indicator apart. Don't attempt this unless you already know how to put it back together! The spring can weaken in one direction over time but the problem is more likely the result of previous repairs. Just the act of taking the mechanism apart can distort the spring's shape. The difference in force shouldn't matter much but if it's imperative to your operation that they are identical, you should consider switching to a different brand. Bestest, Tesatast and Compac would be ideal substitutes.

Q: We have an Interapid 312B-2, the hair spring is currently not wound up, when you're looking at the dial and needle as if you were using the indicator, would you spin the needle clockwise to put tension on the spring or counter clockwise to put tension on the spring? The crown gear has a bent tooth which stops the large needle from turning more than 4 revolutions in the counter clockwise direction. Can the crown gear be removed and turned so the bent tooth doesn't hinder the movement?

⇒ You will have to disassemble the gage in order to pre-load the hairspring. When it is apart, you will see the direction required. The crown gear can also be positioned so that the damaged tooth is not engaged.

Q: I have an Interapid indicator that has a little play side to side with the end. It will repeat fine as long as you don't move side to side on the part you're indicating. I was wondering if you could give me some advice on what i can do to fix the play?

- → You would want to tighten the ball bearings just enough to remove the play. If you over- tighten, your indicator will start to stick or not repeat correctly.

INTERAPID M1.7 CONTACT POINTS

How are they measured? From the shoulder (not including the thread) to the center of the ball. Contact point thread size is M1.7 x 0.35 Those are the measurements given below.

Why do many catalogs have different lengths from those listed below? The Interapid carbide point lengths shown in the catalogs are the length of protrusion from the front of the indicator. A useful figure in some circumstances but not particularly helpful if you have a handful of contact points and you need to sort them out. We have also listed the actual length of the point from the **shoulder** to the center of the ball as a reference.

Usable, exposed catalog length is the amount of protrusion (L) as shown in the sketch. The actual length is about 2 mm longer.

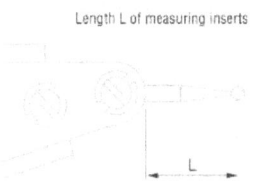

Length L of measuring inserts

Ball diameter is .080" (2 mm) on new Interapid indicators. Carbide is standard with these models.

Most significantly, for accurate readings, Interapid contact points must be at an angle of about 12-15 degrees from the work surface. This is more critical with .0001" indicators where errors can quickly add up.

Catalog length = .650" / Actual length = .687"

For the following models:

- 312B-1
- 312B-2
- 312B-3
- 312B-4
- 312B-20

- 312B-1v
- 312B-2v
- 312B-3v
- 312B-20v

Ball material	Ball Ø	Actual length	Manufacturer's equivalent	Our Part Number
carbide	.015"	.687"	74.116284	MTC39-02
carbide	.031"	.687"	74.105998	MTC39-03
carbide	.060"	.687"	74.105997	MTC39-04
carbide	.080"	.687"	74.105996	MTC39-05
Nylon	.060"	.687"	none	MTC39-19
ruby	.040"	.687"	none	MTC39-13
ruby	.080"	.687"	none	MTC39-14
Teflon	.060"	.687"	none	MTC39-16
Teflon	.080"	.687"	none	MTC39-17

CATALOG LENGTH = 2.675" / ACTUAL LENGTH = 2.750"

For the following models:

- 312B-15
- 312B-15V

Ball material	Ball Ø	Actual length	Manufacturer's equivalent	Our Part Number
carbide	.015"	2.750"	none	MTC40-33
carbide	.031"	2.750"	74.111910	MTC39-34
carbide	.080"	2.750"	74.111913	MTC39-35
carbide	.100"	2.750"	74.111912	MTC39-36

CATALOG LENGTH = 16.5 MM / ACTUAL LENGTH = 18.5 MM

For the following models:

- 312-1
- 312-2
- 312-1v
- 312-2v

Ball material	Ball Ø	Actual length	Manufacturer's equivalent	Our Part Number
carbide	2.0 mm	18.5 mm	74.105993	MTC40-05
carbide	1.5 mm	18.5 mm	74.105994	MTC40-04
carbide	0.8 mm	18.5 mm	74.105995	MTC40-03

CATALOG LENGTH = 15.2 MM / ACTUAL LENGTH = 17.2 MM

For the following models:

- 312-3
- 312-4

Ball material	Ball Ø	Actual length	Manufacturer's equivalent	Our Part Number
carbide	2.0 mm	17.2 mm	74.110482	MTC40-15
carbide	1.5 mm	17.2 mm	74.110491	MTC40-14
carbide	0.8 mm	17.2 mm	74.110507	MTC40-13
Teflon	2.0 mm	17.2 mm	none	MTC40-17

Contact points can be bought online by visiting our website: www.longislandindicator.com

"I like the scientific spirit—the holding off, the being sure but not too sure, the willingness to surrender ideas when the evidence is against them: this is ultimately fine—it always keeps the way beyond open—always gives life, thought, affection, the whole man, a chance to try over again after a mistake—after a wrong guess." — Walt Whitman, WALT WHITMAN'S CAMDEN CONVERSATIONS

DOUBLE LENGTH

You can use these double length points on any of the models listed. The readings on the dial will be **doubled** as a result. The graduations on the .0005" indicator will therefore equal .001" and the graduations on the .0001" indicator will equal .0002". This only works if the contact point is exactly twice the length of the standard point.

For the following models:

- 312B-1
- 312B-2
- 312B-3
- 312B-4
- 312B-20

- 312B-1v
- 312B-2v
- 312B-3v
- 312B-20v

Ball material	Ball Ø	Actual length	Manufacturer's equivalent	Our Part Number
carbide	.030"	1.450"	74.106363	MTC39-33
carbide	.060"	1.450"	none	MTC39-34
carbide	.080"	1.450"	74.106361	MTC39-35
carbide	.120"	1.450"	none	MTC39-37

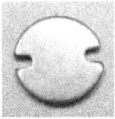

18.60307

Contact point wrench 18.60307 comes supplied with new indicators.

INDEXABLE BENT CONTACT POINT

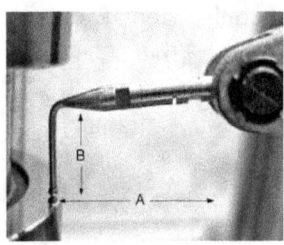

Dimension A and B are approximate. The Indexable bent contact point is designed for use with Interapid test indicators and the photograph gives you an idea of its use.

Any notion of taking direct readings goes right out the window, however. This contortion means that you can only use your indicator for the purpose it was designed: comparison work. If you're not sure whether it'll work for you or not, buy just one piece to test it. We make no guarantees.

Ball diameters are limited and they're only available in carbide.

Length A	Length B	Carbide Ø	Manufacturer's equivalent	Our Part Number
.6875"	.3"	.031"	none	MTC41-01
.6875"	.3"	.060"	none	MTC41-02
1.45"	.3"	.031"	none	MTC41-03
1.45"	.3"	.060"	none	MTC41-04

This custom contact point can be bought online by visiting our website: www.longislandindicator.com

INTERAPID TEST INDICATOR CONTACT POINT INSTALLATION

Make sure your contact point is the correct length for your particular model. Check the section on contact points if you aren't sure.

This page is an excerpt from our *Interapid Test Indicator Repair Manual.*

Here we are comparing it to a new point since it looks like the customer may have installed a generic point. The length matches, and all is well.

Caution! Here you can see a big mistake. The user has screwed the contact point into the rounded side of the pivot. If we do this, we will no longer have an accurate indicator because the contact point will be too long.

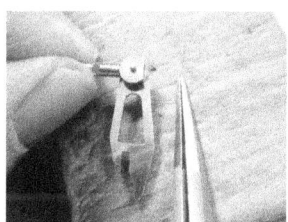

This is wrong! Always make sure the point is screwed into the *flat side of the pivot.*

Check that the contact point is tightly screwed into the flat side of the pivot. Although contact point wrenches are included with Interapid test indicators, we find that our small pliers work just as well.

➡ Note: you do not need to disassemble the indicator to install a new point. Our photo is only designed to illustrate.

MarTest Test Indicator

QUALITY GERMAN TEST INDICATORS

For the connoisseur of German made measuring instruments. MarTest indicators are new model replacements for the popular Puppitast indicator made by Mahr of Germany. Originally patterned after the Swiss BesTest indicators, the new models bear a strong similarity in construction to the older Japanese Mitutoyo models.

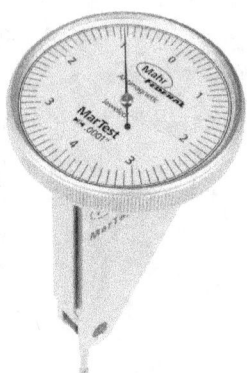

MARTEST VERTICAL MODEL

These indicators are beautiful, sturdy and very responsive.

Test indicators are used for comparative measurements. They can be used in any type of measuring stand. Since the contact point swivels and can be used in the up or down direction, the MarTest indicator is suitable for various measuring and inspection tasks. For example:

- Testing concentricity and run-out of shafts and bores.
- Testing of parallelism and alignment of surfaces.

The major features of the MarTest indicator:

- no slant dial on the MarTest (face lies flat like most other indicators)
- metal bezel turns on a rubber o-ring
- inch models have a blue-green dial face
- metric models have yellow dial face
- contact point angle does not have to be set at 12 degrees (use it at zero degrees to work surface)
- comes with 3/8" mounting shaft on inch models and 8 mm diameter mounting shaft on metric
- has 3 dovetails built into the body
- made in Europe (Czech Republic) to the best of our knowledge
- does not come supplied with holding bar or swivel clamp
- non magnetic

Q: Is the Interapid 18.60307 wrench that you offer interchangeable with the MarTest 4305868 spanner?

➡ The Interapid contact point wrench will not fit the MarTest points. The slot in the wrench is too narrow.

Q: I have a question about using a dead-blow to tap the vise on my mill while I am indicating it with my Mahr 801SG. I used to indicate the vise by running it in the x axis will taping it with a dead-blow while the indicator was on it to dial it in. I no longer do this because I figured the shock is something that may not be good for the DTI. Is this correct or am I being overly cautious?

➡ You are not being overly cautious. Although we can't be certain it will cause any particular damage to the indicator, the vibrations, blows, shocks, bumps, etc. of any kind from any source could be detrimental. You have made the right decision.

MITUTOYO TEST INDICATOR

MULTIPLE REVOLUTION MODELS

Mitutoyo slanted dial test indicator with .080" diameter carbide contact point installed is an excellent choice for all precision measurement.

These popular light weight Japanese imports have the same slanted (tilted) dial which Swiss-made Interapid indicators have been known for since the 1950's. They also make two revolutions in either direction from zero giving them a longer range than the standard indicators.

A serial number and scan image is conveniently printed on the dial face for ISO, inventory, and calibration records control.

The dial hand (pointer) always moves in the clock-wise direction and it has a starting point just to the left of zero; about 11 o'clock. (The Interapid indicators, in comparison, have a hand that is set to 12 o'clock and the hand will move both clockwise and counterclockwise.)

See for yourself if the black dial is something you can deal with. Personally, we think the white dial is easier to read.

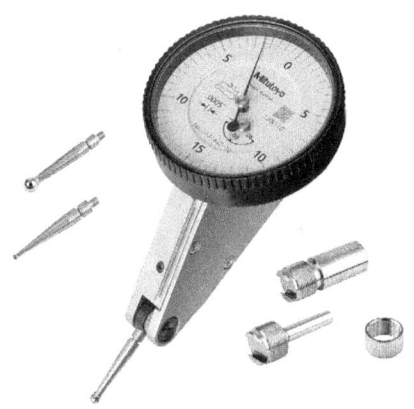

A one-piece plastic bezel and crystal is easy to replace by the end-user when it's damaged. Unlike many other indicators, no special tools will be required. These indicators are accurate and respond very well.

This great starter set Includes the Mitutoyo test indicator with 20° slanted dial plus the following:

- 2 mm diameter (.080") carbide installed
- one 3/8" diameter dovetail stem (removable)
- three dove tails permanently mounted on the body
- box
- certificate of calibration
- dial color: see below
- accuracy: ±.0005" on the .0005" models
- accuracy: ±.0002" on the .0001" models

Graduations	Range	Dial color	Contact point length	Current Model No.
.0005"	.06"	white	.78"	513-442-10E
.0005"	.06"	white	1.33"	513-446-10E
.0001"	.06"	white	.59"	513-443-10E
.0005"	.06"	black	.78"	513-442-16
.0005"	.06"	black	1.33"	513-446-16
.0001"	.06"	black	.59"	513-443-16

What do the Letters mean?

When researching the catalogs or looking online you will usually see the letter E or T at the end of the model number.

Inch **model E** includes

- 3/8" mounting stem
- 2 mm diameter carbide contact point (installed)

Inch **model T** includes

- 3/8" and 4 mm mounting stems
- 1 mm, 2 mm, and 3 mm diameter carbide contact point
- swivel clamp #900322
- holding bar #900306

It is important to note that, unlike the Interapid indicator, these do not require the 13 degree contact point angle when taking measurements. Keep the contact point as close to parallel to the surface of the measured piece as you possibly can.

Can they be repaired?

In a surprising move, Mitutoyo has reduced the number of repair parts to the barest minimum. Only 10 parts make up the entire gage. You can easily replace the crystal since it is part of the bezel assembly that just snaps in place. If anything else is damaged you simply replace the dial movement or the body movement. We no longer have the option of separately replacing hair springs or ball bearings. It is doubtful that repair shops will be able to offer cost-effective repairs, however.

Q: I'm trying to move the contact point to a different angle. Do I just push it or do I have to loosen the nut?

Don't loosen the nut! That's the ball bearing and it has been carefully adjusted by the manufacturer. Use your thumb to move the contact point.

MITUTOYO SINGLE REVOLUTION TEST INDICATOR

NEW MODELS WITH ONE REVOLUTION

One revolution standard test indicators—without the tilted face and dial—are the most popular and best selling indicators in the US. They are also a real bargain when you get right down to it.

New features for 2017: better scratch resistance on crystal, more durable bezel, and the proper contact point length is printed right on the dial. **These new models will have the suffix -10 added to the model number.** But here is our favorite new feature: the indicator consist of just 4 assemblies and if a repair is needed, you can simply buy the entire new movement, new head assembly, crystal or (even) body. You will no longer have to install gears or springs. You will still need some tools, a steady hand and mechanical know-how.

But, because of the low price, they could be considered "throw-away" indicators. When they break down, just toss and replace. Excellent for the factory or shop where they are used by the dozens, if not hundreds.

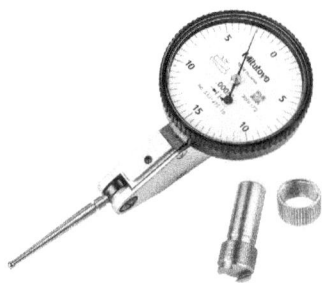

The basic set (E) with a ⅜" stem attachment.

They are also a good bargain when treated with a little bit of respect although they are a poor comparison to their European counterparts, when you start to examine their construction. Yes, the bezel and crystal can be easily replaced but because they are all plastic to begin with, they will break, crack or come undone much more frequently than the sturdier, more permanent designs of other brands. Note: the manufacturer has chosen to use epoxy to keep some parts in place, including the contact point! It may be hard to unscrew your old point if you want to replace it.

- easy to replace crystal and bezel assembly
- excellent repeatability and accuracy
- .0005" models are accurate to ± one graduation
- .0001" models are accurate to ± one graduation
- sturdy, oversized ball bearings are rarely damaged (based on repair reports)
- serial number and scan image printed on the dial face
- basic set includes one 3/8" diameter stem attachment

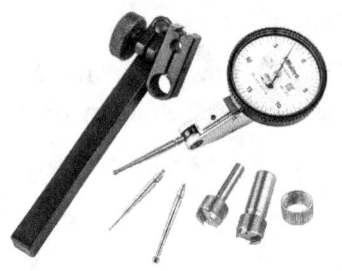

Full set (T) also includes .157" diameter stem attachment, swivel clamp, 4" long holding bar as well as 1 mm and 3 mm diameter carbide ball contact points.

Graduations	Range	Contact point length	Dial	New Model No.	Set
.0005"	.03"	19.9 mm	0-15-0	513-402-10	basic (E)
.0005"	.03"	33.9 mm	0-15-0	513-412-10	basic (E)
.0005"	.03"	19.9 mm	0-15-0	513-402-10	full (T)
.0005"	.03"	33.9 mm	0-15-0	513-412-10	full (T)
.0001"	.008"	15 mm	0-4-0	513-403-10	basic (E)
.0001"	.008"	15 mm	0-4-0	513-403-10	full (T)

For Mitutoyo indicators with dovetails

The **swivel clamp** attaches to the dovetails of all indicators which have dovetails. This includes Starrett, BesTest, Compac, Interapid, Fowler, and even generic Chinese indicators. There are two holes in the clamp, of different diameters, and the clamp is tightened with a single thumb screw.

- Swivel clamp Ø 4 mm and Ø 8 mm holes 900321
- Swivel clamp Ø 5/32" and Ø 3/8" holes 900322
- Holding bar 0.25" x 0.5" x 4" long 900306
- Universal holder Ø 8 mm, 80 mm long 21CZA233 (old #901916).

Stem attachments with different diameters will fit any test indicator that has dovetails whether made by Mitutoyo or someone else. These come with a knurled nut which, when tightened, will firmly clamp the stem to the indicator. The usable length of the stem is 18 mm.

- 4 mm diameter ... 21CZB131 ... $15.00
- 6 mm diameter ... (old 902803) 21CZB128
- 8 mm diameter ... (old 902804) 21CZB129
- 9 mm diameter ... (old 902805) 21CZB130
- knurled nut is not sold separately

What's new in the 2017 models?

It's been 20 years since Mitutoyo updated their test indicators. These will be the models with -10 and -16 suffixes. Here are some noteworthy changes introduced in 2017:

1. Indicators come with a certificate of calibration which is NIST traceable.
2. The spanner nut appears to be recessed but this is because there is now an opening in the cover plate. Once you remove the body cover (4 Phillips screws) you can get at, and remove, the ball bearing spanner without any problems.
3. The cover can now be removed without affecting any internal gears.
4. The plastic underside of the dial assembly is now solid and it is less apt to collect dirt or grime.
5. The dial assembly remains the same.
6. The contact point length is clearly shown on the dial face.
7. The bezel appears to rotate more smoothly.
8. The movement jewels are synthetic sapphire.
9. The plastic bezel-crystal assembly runs on an o-ring and is easily replaced without tools.
10. The bezel-crystal assembly is just slightly taller giving the instrument an overall gawkier look.

The bezels on older models are easily replaced. Use the flat end of a screw driver to pry it off.

MITUTOYO POCKET TEST INDICATOR

POCKET TYPE SERIES 513 – LAST WORD STYLE

The Best of the "Last Word" tubular body type indicators.

These Japanese imports have a structure similar to the well known Starrett Last Word indicator. They have been recently redesigned and the results make them a viable and desirable alternative to the Last Word style indicator. In fact, if you need this particular body style then, among all manufacturers, this is the only model we can recommend.

It will still be necessary to manually switch the contact point direction, using the lever on the side of the body. If this lever is not fully engaged one way or the other, the indicator will not return to zero. If you would like automatic switching, then see the brand comparisons section for some excellent alternative models.

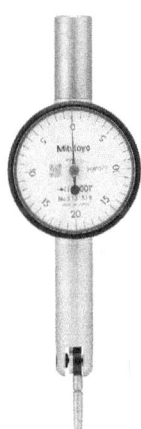

Includes indicator with .078" carbide point installed and one 3/8" holding stem installed. The holding stem can be removed and replaced with different diameters (optional). This type of indicator does not have dove tails. You will mount the indicator by the 3/8" stem. Accessories, other contact points, 4 mm mounting stem, etc., are separately available or you can opt to buy the Full Set "T". Looking at the discounted internet prices below, you will see that sometimes the full set actually costs less than the individual indicator.

Basic model 513-518. This indicator's hand makes one revolution. The dial is balanced (see photo). The contact point can be swiveled more than 90-degrees forwards and backwards. The basic set includes the following:

- Holding stem 3/8" (.375") with wrench
- Plastic box
- Manufacturer's one year warranty

The **FULL SET** (designated by the letter "T" after the model number) is often a real bargain since it also includes six additional accessories:

- .039" diameter carbide contact point
- .118" diameter carbide contact point
- 4 mm diameter holding stem #102036
- swivel clamp #900322 with dovetail, 4 mm and 3/8" diameter clamping holes
- .315" diameter holding rod #900211
- rectangular holding bar 4" long #900306

Graduations	Range	Dial	Contact Point	Model
.001"	.04"	0–20–0	1.04"	513-518
.001"	.04"	0–20–0	1.04"	513-518T
.0005"	.02"	0–10–0	1.47"	513-512
.0005"	.02"	0–10–0	1.47"	513-512T
.0001"	.01"	0–5–0	.74"	513-504
.0001"	.01"	0–5–0	.74"	513-504T

ACCURACY OF POCKET TEST INDICATORS

- 513-518 ±.001"
- 513-512 ±.0005"
- 513-504 ±.0002"

Note: The published **contact point length** is measured from the center of the pivot to the center of the ball. This makes the actual exposed length of the point slightly shorter than listed. For example, the 1.04" contact is only about .9" long once it is screwed in place.

Mitutoyo Test indicator with 2.75" long contact point

Since the Interapid point and the Mitutoyo point both have the same M1.7 thread, you can insert the Interapid point on your Mitutoyo Pocket Style indicators. However, the indicator will NOT read accurately. It you intend to use it only for centering then this makes no difference. Just ignore the graduations on the dial. See the Interapid section of this book for 2.75" long points.

Take note that the Mitutoyo contact points are now epoxied in place. We don't know what possessed Mitutoyo to do this. It may take a bit of force on your part to unscrew the old point and since you will have some epoxy residue in the threads, you may also have to apply a little extra torque to get the Interapid contact point to screw in. Don't hesitate to use a small pair of jeweler's pliers for this purpose. Feel free to ask us to install the contact point on your new indicator.

Mitutoyo indicator repairs and parts

There is not much that can be done by the end user unless special tools and skills are available.

You should never remove the plate on the side of the body. The contact point can be unscrewed using small jeweler's pliers. A wrench is used to change and install the mounting stem. The bezel, still made of durable metal, rides on an o-ring and this bezel can be pried off using a large, flat blade screw driver. The crystal is installed with a crystal press. If you do not have a crystal press it will be nearly impossible to change the crystal. You have the option of buying a new bezel with the crystal already installed (an easy option) or you can send the indicator—or just the bezel—to us for a crystal installation. If you do, be sure to tell us that you only want a new crystal, otherwise we will assume you want a complete refurbish and repair.

While new models can be repaired, it is considered uneconomical to repair the old models such as the popular 513-118 and 513-104 which had unreliable bearings. Also, since these new models are often on sale, the repair cost may not be justified.

FOR "POCKET TYPE" TEST INDICATORS

The **universal holder** for the pocket type indicators such as model 513-518 (and the older 513-118) screws into the end of the indicator. It is not designed for a dovetail. A ball joint allows it to swivel to a desired position.

Threaded mounting stem

- Mounting stems screw directly into the end of the indicator. They come in different diameters and allow for easy, rigid mounting in other holders and magnetic bases.
- rigid mounting stem 4 mm Ø 102036
- rigid mounting stem 3/8" Ø 102081
- rigid mounting stem 8 mm Ø 102822

HOW TO INSTALL A CONTACT POINT

Many test indicators come with a little wrench of some sort designed to help you tighten—or loosen—a contact point. In all cases—including Mitutoyo—small pliers can be used to remove and tighten contact points. Make sure they are nice and snug, hand tightened only. You don't want to force any contact point that refuses to co-operate. Try another point instead. The thread may be damaged, or you may be using a point with the wrong thread. If the thread breaks off inside the indicator, you may have a hard time getting it out and a repair might then be in order.

MITUTOYO M1.7 TEST INDICATOR CONTACT POINTS

GENUINE MTC CONTACT POINTS WITH CARBIDE, RUBY, NYLON OR TEFLON BALLS

Mitutoyo provides two ways of measuring the contact point. The overall length and the more reliable center of ball to shoulder length. We use this second method in the tables on this page. **These points are for current models with M1.7 threads.**

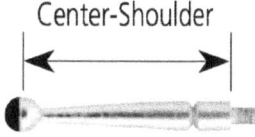

Center-Shoulder

Length is measured from center of ball to shoulder.

Important Note: there was an older, obsolete, model of 513 indicators that had M1.4 threads. Those points are no longer available from us.

Can't find your model number? That's probably because it's no longer being manufactured. If you still have the contact point, just measure it and choose the proper length from the selection below. Older indicators without the suffix (-10 or -16) will take the same size point as the new ones.

CENTER TO SHOULDER LENGTH = 11.2 MM

For the following models:
- 513-405-10
- 513-425-10
- 513-445-10
- 513-455-10
- 513-503
- 513-527

Ball material	Ball Ø	Center to shoulder length	Mitutoyo Equivalent	Our Part Number
carbide	.080"	11.2 mm	103010	MTC53-29

CENTER TO SHOULDER LENGTH = 11.5 MM

For the following models:
- 513-443-10
- 513-443-16
- 513-453-10
- 513-463-10
- 513-473-10

Ball material	Ball Ø	Center to shoulder length	Mitutoyo Equivalent	Our Part Number
carbide	.015"	11.5 mm	—	MTC52-01
carbide	.040"	11.5 mm	136076	MTC52-03
carbide	.080"	11.5 mm	136075	MTC52-05
carbide	.120"	11.5 mm	136077	MTC52-07
nylon	.060"	11.5 mm	—	MTC52-50
Teflon	.060"	11.5 mm	—	MTC52-40
ruby	.080"	11.5 mm	21CZA213	MTC52-60
ruby	.120"	11.5 mm	—	MTC52-61

CENTER TO SHOULDER LENGTH = 15.2 MM

For the following models:
- 513-104
- 513-128
- 513-504
- 513-528

Ball material	Ball Ø	Center to shoulder length	Mitutoyo Equivalent	Our Part Number
carbide	.040"	15.2 mm	131314	MTC51-03
carbide	.080"	15.2 mm	103019	MTC51-05
carbide	.120"	15.2 mm	131315	MTC51-07

CENTER TO SHOULDER LENGTH = .61" (NEW MODELS)

For the following models:
- 513-403-10
- 513-443-10
- 513-443-16
- 513-453-10
- 513-473-10
- 513-463-10

Ball material	Ball Ø	Center to shoulder length	Mitutoyo No.
carbide	.040"	.61"	21CZB065
carbide	.080"	.61"	21CZB064
carbide	.120"	.61"	21CZB066
ruby	.080"	.61"	21CZB112

CENTER TO SHOULDER LENGTH = 16.4 MM (.65")

For the following new models:
- 513-401-10
- 513-402-10
- 513-406-10
- 513-442-10
- 513-442-16
- 513-452-10
- 513-462-10
- 513-472-10
- 513-482-10

Ball material	Ball Ø	Center to shoulder length	Mitutoyo Equivalent	Our Part Number
carbide	.015"	16.4 mm	—	MTC51-14
carbide	.040"	16.4 mm	133196	MTC51-16
carbide	.080"	16.4 mm	133195	MTC51-18
carbide	.120"	16.4 mm	133197	MTC51-20
ruby	.040"	16.4 mm	—	MTC51-60
ruby	.080"	16.4 mm	21CZA204	MTC51-62
ruby	.120"	16.4 mm	—	MTC51-64

CENTER TO SHOULDER LENGTH = 17.4 MM

For the following new models:
- 513-404-10
- 513-444-10
- 513-454-10
- 513-464-10
- 513-474-10
- 513-484-10
- 513-517

Ball material	Ball Ø	Center to shoulder length	Mitutoyo Equivalent	Our Part Number
carbide	.080"	17.4 mm	103006	MTC53-17

CENTER TO SHOULDER LENGTH = 23.0 MM

For the following models:
- 513-118
- 513-518

Ball material	Ball Ø	Center to shoulder length	Mitutoyo Equivalent	Our Part Number
carbide	.040"	23.0 mm	103008	MTC51-29
carbide	.080"	23.0 mm	103007	MTC51-31
carbide	.120"	23.0 mm	103009	MTC51-33

→ Important! On all Mitutoyo test indicators, including the ones with slanted dials, your contact point must move at a 90-degree angle to the measured surface. This also means that the contact point is parallel or 180-degrees with the measured surface.

Contact points can be bought online by visiting our website:
www.longislandindicator.com

CENTER TO SHOULDER LENGTH = 30.4 MM (1.20")

For the following models:
- 513-212
- 513-412-10
- 513-446-10
- 513-446-16

Ball material	Ball Ø	Center to shoulder length	Mitutoyo Equivalent	Our Part Number
carbide	.015"	30.4 mm	—	MTC52-14
carbide	.040"	30.4 mm	136291	MTC52-16
carbide	.080"	30.4 mm	136290	MTC52-18
carbide	.120"	30.4 mm	136292	MTC52-20
ruby	.040"	30.4 mm	—	MTC52-65
ruby	.080"	30.4 mm	21CZA214	MTC52-67
ruby	.120"	30.4 mm	—	MTC52-69
nylon	.060"	30.4 mm	—	MTC52-70
Teflon	.060"	30.4 mm	—	MTC52-80

CENTER TO SHOULDER LENGTH = 33.9 MM

- 513-112
- 513-512

Ball material	Ball Ø	Center to shoulder length	Mitutoyo Equivalent	Our Part Number
carbide	.040"	33.9 mm	131316	MTC52-29
carbide	.080"	33.9 mm	131324	MTC52-31
carbide	.120"	33.9 mm	131317	MTC52-33

STARRETT INDICATORS

New Last Word indicators can be bought from any Starrett distributor and most catalogs carry them. There are numerous variations all having to do with the kinds of attachments that come with the indicators. The ABC part of the Starrett number alludes to these attachments. Darn if we can figure it out but a careful study of their catalog should shed some light on things. We are certain that the letter "Z" means the indicator comes in a storage box and these are usually red in color and have spaces for you to store the other attachments.

The numbering code for Starrett Last Word indicators:

- A = universal shank PT07103A with long and short arm PT07104F
- B = gooseneck shank PT07107A
- C = all attachments included plus 3 steel (chrome plated) contact points
- M = 0.01 mm (metric indicator)
- P = universal friction holder with shank PT13175
- S = most likely indicates that this is a set with attachment(s) included
- Z = storage case
- ? = body clamp PT07101F
- ? = double jointed attachment PT13301
- ? = height gage attachment PT24706 (3/16" x 11/32")
- ? = surface gage attachment PT05119
- ? = coupling with 3/16" hole PT05116

Starrett No.	Graduations	Attachments
711LCSZ	.0005"	included
711FSC	.0005"	with body clamp only
711HSAZ	.0005"	fewer included
711FSAZ	.001"	with magnetic base included
711MGCSZ	0,01 mm	all included

Last Word Accuracy

According to the manufacturer, the accuracy of the Last Word 711 indicator is ±.0005" (though this rather significant detail is omitted in the Starrett catalog).

.001" vs .0005": They're the same indicator but have different dials. The .0005" dial is simply subdivided into smaller units. Both indicators have the same accuracy. You can mentally subdivide the .001" or you can have the dial changed on your old indicator when you send it for repair.

Some notes about using your Last Word indicator

How to change the contact point: swivel the clip, which holds the contact point, using a small screw driver instead of your finger nails. Position the new contact point and swivel the clip back into place. Starrett model 711 Last Word indicators have a contact point which swivels on a ratchet. Sometimes the ratchet doesn't fit well and the contact point will cause the lever arm to jam. You'll have to try another contact point. Sometimes the pivot screw protrudes too much and rubs against the contact point. Again, try another point until one fits. For proper operation make sure the reversing lever is fully engaged up or down and that the contact point is properly seated in the ratchet.

How to change the crystal on the Last Word indicator: remove the chrome bezel by prying it off with a screw driver and replace the crystal. If your old indicator still has a wire spacer, throw the wire away. The new crystals don't need this spacer. A pair of jeweler's pliers will help you squeeze the bezel back into place.

If the indicator hand jumps on a regular basis, try de-magnetizing the indicator. If you don't already have one, a small demagnetizer can be bought from most supply catalogs. Spinning objects and motors can induce magnetic fields in the indicator. If magnetic fields are a problem in your shop environment it may be best to switch to a non-magnetic indicator. Mitutoyo, Bestest, Compac and Interapid are all suitable alternatives.

If the indicator hand hangs up on occasion, just tap the indicator and the hand will probably return to its normal setting. Don't let this upset you if it happens only occasionally. We're talking Starrett here.

Repairs of the Last Word 711 indicator can be surprisingly tricky because new parts don't always fit as well as they should. You should be prepared to make minor alterations from time to time.

Starrett Last Word indicators have problems which are as unique as their design.

If you encounter a hand that sticks on rare occasions and refuses to budge, try tapping the indicator with a screw driver or bang it lightly on the surface of a table. The hand will probably dislodge and go back to normal. This can happen when you have moved the contact point too quickly. If the hand always skips at a certain spot, you will need to have it cleaned or repaired.

The end screw should be in place if you do not have a stem or attachment installed.

Last Word contact points do not screw in place. They have a ratchet that fits under the clip.

How old is my Last Word indicator?

The exact age can't be determined because Last Word indicators have changed very little over the years. They did, however, go through several stages with distinct characteristics. Among the oldest models you will find that the end of the body was slanted (see photo). This is probably pre-1960. Subsequent models have ends which are squared off.

The oldest models also had dials which lacked the yellow band along the right hand side, as seen in current models. You will see .001" written as 1/1000" on the dials (see photo). Keep in mind, your indicator may have been repaired at some time and the dial may have been changed to the new style.

Q: I have a nice, functional .0001" Starrett 711-T1 (which is apparently obsolete now) and the bezel is a little loose; it rocks out of the plane of the dial. There are three, small flathead screws evenly distributed around the bezel. Will gently tightening these tighten up the bezel? If not, is there an alternative means of doing that?

> ➡ The three screws you mention will have no effect on the rocking. The mounting plate has come loose and those screws can only be accessed by removing the bezel, the hand and the dial.

STARRETT 708A - 709A - 708B - 709B TEST INDICATOR

"American Made" light weight test indicators by one of America's original gage manufacturers.

Do not confuse these with the lower-cost Starrett models (3808A for example) which are made in China. The Chinese versions are decidedly inferior in every way. If there is no country of origin printed on the dial, then they are made in China.

With an easy to read tilted dial similar to Swiss made Interapid indicators. The contact point angle must be held at a 15° angle to the work surface for accurate reading. This feature sets it apart from all other test indicators (except Interapid) and can prove very useful. If your shop uses other brands of indicators as well, be sure to instruct everyone to heed the 15-degree angle requirement. Plastic bezel and crystal ride on a rubber o-ring making these easy to remove and replace without special tools. They can be bought as replacement parts directly from Starrett or from Starrett distributors. Three dovetails can accept any mounting stem or holder even if they are not Starrett. Contact point length can be adjusted for calibration with a simple set screw although this is not needed on new indicators. The right half of the dial is shaded with a yellow-orange stripe for use with a mirrored set-up.

Indicators come supplied with:

- 2 mm diameter carbide contact point
- custom fitted box which will also accommodate attachments
- attachments included for some models (see chart below)
- certificate of calibration not included
- single or multiple revolutions (see chart below)
- assorted dial colors for some models (see chart below)

Attachments (included with certain models, see chart below):

- dovetail body clamp
- tool post holder
- swivel post snug with dovetail indicator clamp snug and rod unit

www.longislandindicator.com

B709BCZ comes with attachments.

Range	Graduations	Point Length	Attachments	Dial Color	Model No.	Starrett No.
.010"	.0001"	20 mm	none	white	708A	708AZ
.010"	.0001"	20 mm	none	red	708A	R708AZ
.010"	.0001"	20 mm	none	black	708A	B708AZ
.010"	.0001"	20 mm	yes	white	708A	708ACZ
.010"	.0001"	20 mm	yes	red	708A	R708ACZ
.010"	.0001"	20 mm	yes	black	708A	B708ACZ
.020"	.0001"	20 mm	none	white	708B	708BZ
.020"	.0001"	20 mm	yes	white	708B	708BCZ
.030"	.0005"	20 mm	none	white	709A	709AZ
.030"	.0005"	20 mm	none	red	709A	R709AZ
.030"	.0005"	20 mm	none	black	709A	B709AZ
.030"	.0005"	20 mm	yes	white	709A	709ACZ
.030"	.0005"	20 mm	yes	red	709A	R709ACZ
.030"	.0005"	20 mm	yes	black	709A	B709ACZ
.050"	.0005"	34 mm	none	white	709A	709ALZ
.050"	.0005"	34 mm	yes	white	709A	709ALCZ
.060"	.0005"	20 mm	none	white	709B	709BZ
.060"	.0005"	20 mm	yes	white	709B	709BCZ

STARRETT CONTACT POINTS FOR 708A – 708B – 709A – 709B

- These contact points have 1-64 threads
- They will not fit Starrett indicators made in China

Length	Diameter	Ball Material	Order No.	Starrett No.
20 mm	.015"	carbide	MTC58-01	
20 mm	.031"	carbide	MTC58-02	
20 mm	.080"	carbide	MTC58-04	PT23914

Easy to read tilted dial similar to Swiss made Interapid indicators. The contact point angle must be held at a 15° angle to the work surface for accurate reading.

Contact points can be bought online by visiting our website:
www.longislandindicator.com

Accuracy

According to the Starrett web site, these gages "meet or exceed ISO accuracy requirements" but they do not specify or publish what those accuracies are.

Repairs

The Starrett 708 and 709 series of test indicators have some idiosyncrasies that can make these repairs challenging. Unless you have experience and tools, we strongly urge you to leave this in the hands of professionals. Only the clear plastic cover (crystal and bezel assembly) and the contact points are easily replaced by the end-user.

Spare parts are not available from us because we are not Starrett distributors. If you can not find a distributor who is willing to sell parts (there is little money to be made selling parts) then you can contact Starrett directly. Check their web site for a parts breakdown and contact information.

It is important to remember that these Starrett indicators are accurate when the contact point has an angle of about 10 to 15 degrees to the test surface. If, during calibration, you find that you are losing or gaining distance (even though the contact angle is correct) you will need to shorten or lengthen the contact point. The longer the point is, the less travel it will register. To do this, you will have to adjust the set screw which you will see on the opposite side of the pivot. Loosen the contact point first and then adjust the set screw. Now you can tighten the contact point again. Quite a bit of trial and error will get you the results you're after. If this is too much of a headache, next time don't buy Starrett.

Repair hint

If you need to remove the guts from the body you will have to remove adjusting screw #8 first (refer to online parts diagram.) Similarly, if you need to remove the dial head assembly #11 you will have to remove (or at least loosen significantly) adjusting screw #36 on the cover side. You can only find this screw by removing the cover. A very small jeweler's screw driver will be required.

Q: I am hoping that you can help me out with this. One of our guys picked up a used Starrett indicator. We are trying to get the original contact for it. I ordered the MTC58-04 from your site. From what I could tell this should have been the right one but the threads are different. Would you be able to tell me what tip it needs?

→ Well… here's the low-down. Take a look at the Starrett dial and you will see one glaring omission: Nowhere does it say "American Made" or "made in USA". Guess what? You are the owner of a Chinese indicator! The best approach is to contact Starrett to see if they sell replacement points (which I suspect they do). I would check their web site as a starting point.

"Science, my lad, is made up of mistakes, but they are mistakes which it is useful to make, because they lead little by little to the truth." — Jules Verne, A JOURNEY TO THE CENTER OF THE EARTH

TESATAST TEST INDICATOR

The ideal test indicator when working in a metric environment. Many of the Tesatast models are unavailable in the U.S. but, if you find them, they are worth the investment.

High Quality Swiss test indicators with **metric** attachments. For all intents and purposes, the Swiss-made Tesatast indicator is identical to Bestest, Swisstast, and Roctest. All parts are interchangeable, including contact points.

WHAT IS THE DIFFERENCE BETWEEN TESATAST AND BESTEST?

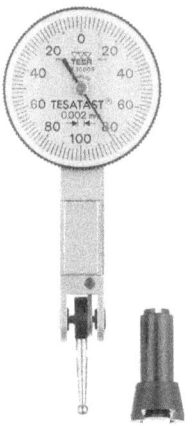

Tesatast and BesTest are both made by the same manufacturer (TESA) and only the dials and certain accessories differ. Most significantly, Tesatast indicators come with a metric 8mm diameter dovetail extension attachment, while Bestest is supplied with 1/4" diameter extensions. You will also find Tesatast has several configurations, particularly those with long contact points (1.5"), which are not available in the Bestest models. Conversely, if you don't find the configuration you need here, be sure to look at the Bestest models. All the necessary accessories and contact points are available.

Supplied with 8mm diameter dovetail extension and case. This is the perfect metric replacement for popular BesTest indicators. Like their twin, these are light-weight and responsive.

TESATAST HORIZONTAL METRIC TEST INDICATOR

Accuracy of the Tesatast models:

- 0.01 mm ± 0.01 mm
- 0.002 mm ± 0.002 mm

Certificate of Accuracy is provided without test data. The certificate guarantees the accuracy as shown above. If you need full certification, you will have to send the indicator to a calibration lab.

Graduations	Range	Dial Ø	Contact Point	TESA Model No.
0,01 mm	0,8 mm	28 mm	12,5 mm	18.10005
0,01 mm	0,8 mm	38 mm	12,5 mm	18.10006
0,002 mm	0,2 mm	28 mm	12,5 mm	18.10009
0,002 mm	0,2 mm	38 mm	12,5 mm	18.10010
0,01 mm	0,5 mm	28 mm	36,5 mm	18.10007
0,01 mm	0,5 mm	38 mm	36,5 mm	18.10008

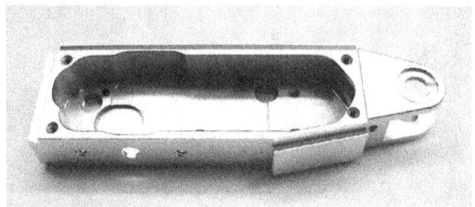

Tesatast and BesTest use identical parts including contact points.

Tesatast Vertical Test Indicator

Supplied with 8mm diameter dovetail extension and case. This is the perfect metric replacement for popular BesTest indicators.

Accuracy of the Tesatast models:

- 0.01 mm ± 0.01 mm
- 0.002 mm ± 0.002 mm

Graduations	Range	Dial Ø	Contact Point	TESA Model No.
.0005"	.030"	1"	.5"	18.20204
.0005"	.020"	1.5"	1.5"	18.20205
0,01 mm	0,8 mm	28 mm	12,5 mm	18.10204
0,01 mm	0,5 mm	28 mm	36,5 mm	18.10205
0,002 mm	0,2 mm	38 mm	12,5 mm	18.10304

The most common reasons why these indicators come in for repair:

1. The bezel is dented and the crystal is cracked or has popped out.
2. Oil has gotten into the hair spring and the indicator no longer repeats or is sluggish.
3. The teeth on the crown gear are damaged due to shock.
4. The ball bearings are ruined due to shock at the contact point end.
5. The contact point has broken off leaving the thread in the pivot.
6. Everything is "gummed up" with coolant or grime.
7. Someone has tampered with ball bearing adjustment.

TESATAST PARALLEL TEST INDICATOR

Parallel style Tesatast models have dials mounted on the side of the body. There are 3 separate dovetails along the edges of the body. Supplied with 8mm diameter dovetail extension and case. This is the perfect metric replacement for popular BesTest indicators.

Why are these "currently unavailable" metric models being featured on this page? We can hope that the metric models will be put back on the market at some point. Quite possibly they are available in Europe where metric manufacturing is the norm.

Q: Could I order a new crystal for this indicator [shown above]? I know that it looks beat-up, but still works fine. If a new or used bezel is also available and not too expensive, please also tell me how much that would be?

→ It will not be possible to install a new crystal in a bezel which is no longer perfectly round. Replacing the bezel is not easy. Unfortunately it is time to send it to a repair shop.

Q: The machinist who is requesting this Tesatast indicator was wondering if there was any was to make it more oil resistant. That seems to be the main issue with many of our indicators and due to the cost of this indicator it would help extend its life.

→ None of the test indicators are immune to oil or liquid contamination. Although not as much of a problem if the oil sprays from above, it can quite obviously enter the body near the contact point. Some people have experimented with plexiglass shields. Perhaps this is something you can look into.

Q: I have a side mounted dial Tesatast indicator (18.10011) and one of the screws securing the dial is loose. Consequently the dial is not moving freely if not completely centered. Are you able to advise as to how I can remove the crystal and dial to secure the screw?

→ This is not an easy process to describe. Please refer to our Repair Manual.

TEST INDICATOR HOLDERS

Although specifically designed for Bestest, Interapid, Tesatast indicators, these holders and attachments will work with any manufacturer's indicator that has a dovetail.

BROWN & SHARPE UNIVERSAL DOVETAIL ATTACHMENT

Brown & Sharpe 599-7054 universal dovetail attachment fits any test indicator with dovetails such as BesTest, Interapid, and Mitutoyo. The 1/4" diameter shank swivels until the knurled nut is tightened. Length of shank is 2-1/2" from the end to the center of the ball.

B&S 599-7054

DOVETAIL SHANK (FIXED STEM)

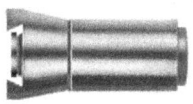

These **non-swiveling** stems attach to any dovetail on any brand test indicator. There's a screw slot on the end of the stem. Turning this with a standard screwdriver will tighten the stem onto the dovetail. Swiss-made by TESA for the Bes-Test indictor.

Order No.	Dimensions
01840104 (18.40104)	Ø 4 mm x 12 mm
01840105 (18.40105)	Ø 8 mm x 12 mm
599-7052 (18.50104)	Ø 7/32" x .5"
599-7053 (18.50105)	Ø 1/4" x .5"

Angle Holder

Cylindrical shank allows test indicators to be held at right angles. The thumb screw tightens down onto an 8mm or 3/8" stem attachment on your indicator. Made in Switzerland by TESA.

- 8mm diameter x 25mm long (bore = 8mm) 01840406 (18.40406)
- 3/8" diameter x 1" long (bore = 3/8") B&S 599-7043 (TESA 18.50406)

Rectangular or Cylindrical Shank

This indispensable universal holder for lathes and height gages has been newly redesigned by the manufacturer. They replace the former Compac model SP and SPA holder. The rectangular or cylindrical shank attaches firmly to any dovetail (any manufacturer) and then clamps into the tool post holder or into the height gage. The indicator can swivel into many positions and then stays securely fixed when the knob is tightened. It will also accommodate a dial indicator with a

standard 3/8" diameter stem. It's also an ideal way to calibrate both your test and dial test indicators when used with a height gage and some gage blocks. Order one shank (rectangular or cylindrical) and one clamp shown below. Swiss-made by TESA.

Works in conjunction with a clamp. Order the items separately from the list below. The clamp attaches to any dovetail. The rectangular or cylindrical shank fits into the 5.6mm diameter hole on the clamp (shown below). Tightening the knob will tighten both the rectangular shank and the dovetail at once.

Order No.	Type	Dimensions
01850203 (599-7047)	rectangular	1/2" x 1/4" x 3" long
01850202 (18.50202)	cylindrical	Ø 3/8" x 3.5" long

TEST INDICATOR HOLDING CLAMP

Test indicator clamp with small hole = 5.6mm for rectangular and cylindrical shanks (shown above) and large hole = 3/8" for indicator. Will also fit the dovetails on any manufacturer's indicators. Can be disassembled into components.

- TESA number 01860401 (B&S number: 599-7045-1)

INTERAPID INDICATOR SWIVEL HOLDER

This versatile Swiss-made mounting stem (axial support) shown in the diagram above is 5-1/4" long and swivels at two points. The two joints allow for multiple positioning. One end has the dovetail attachment and an opening for the Interapid 4mm stem. The other end has a mounting rod which can be gotten in two different diameters as follows:

- 3/8" Ø rod (74.106931) 074106931
- 8 mm Ø rod (74.106026) 074106026

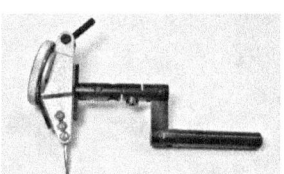

A note of interest: there is a limited swivel of about 20% between the two leftmost joints on this holder. If you're at all handy with a Dremel and an abrasive sanding drum, you can modify the corners of these joints to give you complete side-to-side swivel.

STEM ADAPTER FOR 4 MM STEMS

Interapid indicators come with a 4 mm diameter stem attached to the end of the instrument. This elegant Swiss-made sleeve, shown in the photo above, will fit over the stem so that you can attain different diameters. You can use this adapter on any 4 mm stem–Interapid or otherwise–to increase its diameter. The choices are:

- 3/8" adapter (74.108943) 074108943
- 8 mm adapter (74.108942) 074108942

Test Indicator Contact Points

WHICH POINT TO CHOOSE?

There are four components to take into consideration:

- ball diameter
- thread diameter
- contact point length
- ball composition

New indicators come supplied with .080" diameter balls as standard, but for special needs other diameters are often available. Thread diameters vary from manufacturer to manufacturer but are generally metric threads. Manufacturers differ in their method of measuring contact point lengths. It is imperative that the correct length point is used.

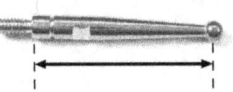

The contact point length is usually measured from the center of the ball to the edge of the shoulder.

WHAT ARE THE POINTS MADE OF?

Our MTC contact points are turned from #303 stainless steel and have carbide, ruby, Nylon or Teflon balls.

Alloy 303 is a non-magnetic, austenitic stainless steel that cannot be hardened by heat treatment. It is the "free machining modification of the basic 18% chromium / 8% nickel stainless steel." [source: pennstainless.com]

We recommend carbide balls over chrome or steel for its durability. For special applications you may need the ruby, sapphire, teflon, nylon or some other exotic.

The contact points listed here are home grown. They're every bit as good as the originals yet they cost less. We give you the manufacturer's ordering number for ease of reference.

How to Install a Contact Point

Most test indicators come with a wrench of some sort designed to help you tighten—or loosen—a contact point. Some points have a little hole drilled through them so that you can insert a pin which can then be used to twist the point. Others have two flat spots designed to work with the wrenches supplied. Still others have a hexagonal area which works very well with a pair or small pliers. In all cases, small pliers can be used to remove and tighten contact points. Make sure they are nice and snug, hand tightened only. You don't want to force any contact point that refuses to co-operate. Try another point instead. The thread may be damaged, or you may be using a point with the wrong thread. If the thread breaks off inside the indicator, you may have a hard time getting it out and a repair might then be in order.

Ruby, Nylon and Teflon points

These contact points are used for special applications, typically in optics. The ruby point outlasts standard chrome points and does not expand or contract with heat and cold. It does not conduct an electric current so it can be used on your machines without creating interference. It is also less likely to scratch delicate surfaces. It's also the material that will hold up to silicon carbide surfaces. The Teflon point is softest made of a solid Teflon ball. It is very white in appearance. It is least likely to scratch but is not as long lasting as the others. The smallest Teflon diameter available is 1/16" (.062"). Smaller points can not be manufactured in this material. The Nylon point is not as soft as Teflon but will last longer. It can also be used where surface damage is an issue. Nylon appears opaque.

Cosine Error

We know that the length of the contact point determines the ratio of leverage. In order to get accurate dial readings the length must be as specified by the manufacturer.

Changing the angle of the contact point changes the ratio just as if you had installed a somewhat shorter point. As a result, the readings on the dial will be higher than they should be.

To compensate for this cosine error it will be necessary to multiply the reading on the dial by the cosine of the angle between the contact point and the measuring surface.

How do you measure the angle? Most of us can intuitively judge an angle of 45° and, perhaps with a little less accuracy, an angle of 30°. Beyond that, we need help. A simple protractor like we used in grade school will suffice. Lay the straight edge on the measuring surface and, by eye, make a judgement of the angle. A few degrees one way or another won't be significant.

Where do you find the cosine? Use a scientific calculator or simply ask Siri.

Enter the value of the angle: for 20° enter "20" and press the COS (cosine) button. The display will read .9340 (allowing for rounding up).

Now it's simply a matter of multiplying the reading on your test indicator with this value.

- For example: when the reading is .0006" you multiply .0006" x .9340 and the result is .00056"
- This is just as expected. Your reading of .0006" was larger by .00004" than it should have been.

From this example you can see that angles of 20 degrees or less have relatively insignificant effects on the reading. But, when the angles are larger you'll be wise to make cosine error adjustments.

This same cosine error applies to all makes of test indicators regardless of contact point lengths and resolutions. Some Fowler test indicators feature "pear-shaped" contact points which compensate, in theory, for the cosine error. These may be a good choice if you frequently use a variety of measuring angles.

Interapid indicators pose a special problem. They're designed to work without cosine error at 12°. When setting up, use a protractor to make sure you're in the ball park. If you're off, then you'll have to compensate, just like the others. If the angle is 32° then the cosine will have to be of the difference between 32° and 12°, in other words 20°. Make the calculations as above.

Longer Contact Points

You may wish to substitute a long contact point on short point models. If the point is exactly twice the length then the readings on the dial will be doubled. A .0001" indicator will now read .0002" and a .0005" indicator will now read .001".

The indicators were not designed with these long points in mind and you may see some problems with repeatability and hysteresis because these long points weigh more. The extra weight interferes with the hair spring and return spring and that may be noticeable in .0001" models.

Once this long point is installed, be sure to label the indicator with the new readings so no one else will make mistakes.

"First we thought the PC was a calculator. Then we found out how to turn numbers into letters with ASCII — and we thought it was a typewriter. Then we discovered graphics, and we thought it was a television. With the World Wide Web, we've realized it's a brochure." — Douglas Adams

INDICATOR MODEL EQUIVALENTS

This section is designed to help you find equivalents to inch reading test indicators which you might own or want to replace. It will show the new models when others have become obsolete. It can also help you find better quality equivalents or—vice versa—cheaper alternatives.

We've taken the major parameters into account: the range, the graduations, the dial diameter and the contact point length. There will of course be some variation from one manufacturer to another. We've rounded the dial and point lengths to the nearest .1" to make comparison easier.

HORIZONTAL INCH MODEL EQUIVALENTS

Where they are made: it is quite likely that most of the manufacturers use at least some Chinese components. Switzerland is quite strict with its labeling requirements and "Swiss Made" can only be used if final assembly plus the majority of the components are of Swiss origin. Other countries may or may not have similar regulations.

We have included Brand name models that are generally available in the U.S. We make no claim that this list is complete. Checking the manufacturer's catalog is the best way to verify the accuracy of our information

- Alina ... Switzerland
- BesTest ... China (most models prior to 2018 were labeled "Swiss made")
- Enco ... Japan unless otherwise noted
- Federal ... various (Switzerland, England, USA)
- Fowler ... China unless otherwise noted
- Gem ... USA
- Girod ... Switzerland
- Interapid ... Switzerland
- Kafer ... Germany
- MarTest ... Germany
- Mitutoyo ... Japan
- Peacock ... Japan
- SPI ... China (current models)
- Starrett ... USA unless otherwise noted (China)
- Teclock ... Japan
- Tesatast ... Switzerland

Grads	Range	Dial Ø	Point Length	Brand	Model No.	Notes
.00005"	.008"	1.5"	.5"	B&S BesTest	7033-3	
.00005"	.008"	1.5"	.5"	B&S BesTest	7033-5	black dial
.00005"	.008"	1.5"	.5"	Tesatast	18.20013	
.00005"	.008"	1.5"	.5"	MarTest	801 SGE	

Grads	Range	Dial Ø	Point Length	Brand	Model No.	Notes
.0001"	.008"	1"	.5"	B&S BesTest	7032-3	
.0001"	.008"	1"	.5"	B&S BesTest	7032-5	black dial
.0001"	.008"	1"	.5"	Fowler (Girod)	52-562-253	
.0001"	.008"	1"	.5"	Tesatast	18.20011	
.0001"	.008"	1.1"	.6"	Mitutoyo	513-463-10	short body
.0001"	.008"	1.25"	(?)	Starrett	3808A	made in China
.0001"	.008"	1.5"	.5"	Fowler UltraTast	52-568-015	made by Käfer
.0001"	.008"	1.5"	.5"	B&S BesTest	7023-3	
.0001"	.008"	1.5"	.5"	Fowler (Girod)	52-562-453	
.0001"	.008"	1.5"	.5"	Tesatast	18.20012	
.0001"	.008"	1.5"	.5"	MarTest	801 SGM	
.0001"	.008"	1.5"	.6"	Mitutoyo	513-403-10	
.0001"	.008"	1.6"	.5"	Mahr Puppitast	800 SMZ	
.0001"	.008"	1.6"	(?)	Starrett	3908A	made in China
.0001"	.01"	1.4"	.8"	Starrett	708A	

Grads	Range	Dial Ø	Point Length	Brand	Model No.	Notes
.0001"	.01"	1.5"	.4"	Teclock	LTI-370	
.0001"	.01"	1.5"	.6"	Teclock	LTI-355	
.0001"	.016"	1.5"	.6"	Mitutoyo	513-443-10	
.0001"	.016"	1.5"	.7"	Interapid	312B-3	
.0001"	.016"	1.5"	.6"	MarTest	801 SRM	
.0001"	.02"	1.4"	.8"	Starrett	708B	

Grads	Range	Dial Ø	Point Length	Brand	Model No.	Notes
.0005"	.03"	1"	.5"	B&S BesTest	7030-3	
.0005"	.03"	1"	.5"	B&S BesTest	7030-5	black dial
.0005"	.03"	1"	.5"	Tesatast	18.20006	
.0005"	.03"	1"	0.3"	Starrett	711H	Last Word Type
.0005"	.03"	1.1"	0.5"	Fowler UltraTast	52-568-005	made by Käfer
.0005"	.03"	1.1"	0.5"	MarTest	801 S	
.0005"	.03"	1.1"	0.8"	Mitutoyo	513-462	
.0005"	.03"	1.1"	1.4"	Fowler UltraTast	51-568-022	made by Käfer
.0005"	.03"	1.1"	1.5"	Mahr Puppitast	800 SLZ	
.0005"	.03"	1.2"	.6"	Fowler	52-562-778	black dial
.0005"	.03"	1.2"	(?)	Starrett	3809A	made in China
.0005"	.03"	1.4"	.4"	Gem	335-30	Last Word Type
.0005"	.03"	1.4"	.8"	Starrett	709A	
.0005"	.03"	1.4"	.8"	Teclock	LTI-315	
.0005"	.03"	1.4"	.8"	Starrett	811-5	head swivels

.0005"	.03"	1.4"	.8"	Teclock	LTI-352	
.0005"	.03"	1.4"	.8"	Teclock	LTI-352B	black dial
.0005"	.03"	1.4"	1.5"	Teclock	LTI-353	
.0005"	.03"	1.5"	.5"	Fowler UltraTast	52-568-010	made by Käfer
.0005"	.03"	1.5"	.5"	MarTest	801 SG	
.0005"	.03"	1.5"	.5"	BesTest	7031-3	
.0005"	.03"	1.5"	.5"	BesTest	7031-5	black dial
.0005"	.03"	1.5"	.5"	Tesatast	18.20007	
.0005"	.03"	1.5"	.6"	MarTest	801 SG	
.0005"	.03"	1.5"	(?)	Mitutoyo	513-302G	swivels 360°
.0005"	.03"	1.5"	.8"	Mitutoyo	513-402-10	
.0005"	.03"	1.5"	.8"	Mitutoyo	513-472-10	ruby contact
.0005"	.03"	1.5"	1.3"	Mitutoyo	513-412-10	
.0005"	.03"	1.8"	.75"	Starrett	3909A	made in China
.0005"	.04"	1.3"	.7"	Teclock	LTI-316	
.0005"	.05"	1.4"	1.4"	Starrett	709AL	
.0005"	.06"	1"	.7"	Interapid	312B-2	74.111371
.0005"	.06"	1"	.8"	Fowler (Girod)	52-562-252	
.0005"	.06"	1.4"	.8"	Starrett	709B	
.0005"	.06"	1.5"	.6"	MarTest	801 SR	
.0005"	.06"	1.5"	.7"	Interapid	312B-1	74.111370
.0005"	.06"	1.5"	1.3"	Mitutoyo	513-446-10	
.0005"	.06"	1.5"	2.8"	Interapid	312B-15	

Grads	Range	Dial Ø	Point Length	Brand	Model No.	Notes
.001"	.03"	1"	.5"	B&S BesTest	7029-3	
.001"	.03"	1"	.5"	B&S BesTest	7029-5	black dial
.001"	.03"	1"	.5"	Tesatast	18.20001	
.001"	.03"	1"	.3"	Starrett	711F	Last Word
.001"	.03"	1"	.3"	Gem	222	Last Word Type
.001"	.03"	1.1"	.6"	Teclock	LTI-310	
.001"	.04"	1.1"	.7"	Mitutoyo Mini	513-528	short body
.001"	.04"	1.3"	1"	Mitutoyo Pocket	513-518	Last Word Type
.001"	.06"	1"	.7"	Interapid	312B-20	
.001"	.06"	1.4"	1.3"	Starrett	811-1	dial tilts

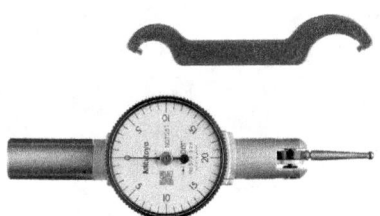

Mitutoyo Mini 513-528 has a shorter body.

"I find television very educating. Every time somebody turns on the set, I go into the other room and read a book." — Groucho Marx

VERTICAL TEST INDICATOR INCH MODEL EQUIVALENTS

This page is designed to help you find equivalents to test indicators which you might own or want to replace. It will show the new models when others have become obsolete. It can also help you find better quality equivalents or—vice versa—cheaper alternatives.

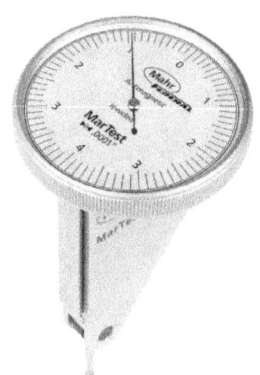

These are the end mounted, vertical, and jig bore style indicators.

We have taken the major parameters into account: the range, the graduations, the dial diameter and the contact point length. There will of course be some variation from one manufacturer to another. We've rounded the dial and point lengths to the nearest .1" to make comparison easier.

This listing does not pretend to be complete. We have only included quality brand names generally available in the U.S.

Grads	Range	Dial Ø	Point L	Brand	Model	Notes
.00005"	.008"	1.5"	.5"	B&S BesTest	7038-3	
.0001"	.008"	1"	.5"	Fowler Girod-Tast	52-562-273	
.0001"	.008"	1.5"	.5"	B&S BesTest	7024-3	
.0001"	.008"	1.5"	.5"	Tesatast	18.20304	
.0001"	.008"	1.5"	.5"	Fowler Girod-Tast	52-562-473	
.0001"	.008"	1.5"	.6"	Mitutoyo	513-453-10	
.0001"	.016"	1.5"	.7"	Interapid	312B-3V	74.111957
.0005"	.02"	1.5"	1.4"	Tesatast	18.20205	

.0005"	.02"	1.5"	1.5"	Fowler Girod-Tast	52-562-474	
.0005"	.03"	1"	.5"	B&S BesTest	7037-3	
.0005"	.03"	1"	.5"	Fowler Girod-Tast	52-562-272	
.0005"	.03"	1"	.5"	Tesatast	18.20204	
.0005"	.03"	1.1"	.5"	MarTest	801 V	
.0005"	.03"	1.1"	.5"	MarTest	801 V	
.0005"	.03"	1.5"	.8"	Mitutoyo	513-452-10	
.0005"	.06"	1"	.7"	Interapid	312B-2V	74.111378
.0005"	.06"	1.5"	.7"	Interapid	312B-1V	74.111375
.0005"	.06"	1.5"	2.8"	Interapid	312B-15V	74.111376
.001"	.06"	1.2"	.7"	Interapid	312B-20V	74.111379

Interapid indicators have two model numbers. (See above)

"I'm sorry to say that the subject I most disliked was mathematics. I have thought about it. I think the reason was that mathematics leaves no room for argument. If you made a mistake, that was all there was to it." — Malcolm X, THE AUTOBIOGRAPHY

SIDE MOUNTED PARALLEL STYLE TEST INDICATOR

Parallel style (lateral model) test indicators are used in specific applications where mounting the standard indicators is impractical. They function just like the regular models but have the dial mounted on the side of the body. Because there is little demand for this style, it may not always be in stock. Here we list a selection of quality indicators that we can recommend.

WHERE THEY ARE MADE:

- BesTest (nominally: Switzerland)
- MarTest — Mahr Federal — no country of origin identification
- Mitutoyo (Japan)
- TesaTast (Switzerland)

GRADS	Range	Dial Ø	Point L	Brand	Model Details
.0005"	.030"	1.1"	.5"	TesaTast	18.20014
.0005"	.030"	1.1"	1.5"	BesTest	599-7021-3
.0005"	.030"	1.1"	14,5 mm	MarTest	801H (4303950)
.0005"	.030"	1.6"	.8"	Mitutoyo	513-282GT
.00005"	.008"	1.1"	1.5"	BesTest	599-7022-3
0,01 mm	0,8 mm	40 mm	21 mm	Mitutoyo	513-284GT
0,01 mm	0,8 mm	28 mm	12,5 mm	TesaTast	18.10011
0,02 mm	2 mm	38 mm	36,5 mm	TesaTast	18.10012
0,002 mm	0,2 mm	28 mm	12,5 mm	TesaTast	18.10013

METRIC TEST INDICATOR MODEL EQUIVALENTS

This page is designed to help you find equivalents to test indicators which you might own or want to replace. It will show the new models when others have become obsolete. It can also help you find better quality equivalents or—vice versa—cheaper alternatives. However, you'll note that not all manufacturers have been represented.

We've taken the major parameters into account: the range, the graduations, the dial diameter and the contact point length. There will of course be some variation from one manufacturer to another. Mitutoyo contact points are actually somewhat shorter than listed, for example. We've rounded off the dial and point lengths to make comparison easier.

Where are they made?

- BesTest (Brown & Sharpe, China)
- Fowler Girod-Tast (Girod) - Switzerland
- Interapid (Tesa) - Switzerland
- Kafer - Germany
- MarTest - Germany
- Mitutoyo - Japan
- Starrett - USA
- Teclock - Japan
- Tesatast (Tesa) - Switzerland

HORIZONTAL METRIC MODELS

Graduations	Range	Dial Ø	Point L	Brand	Model
0,01	0,5	28	14,5	MarTest	800 SA
0,01	0,5	28	18,7	Mitutoyo	513-466-10E
0,01	0,5	28	36,5	Tesatast	18.10007
0,01	0,5	28	41,2	MarTest	800 SL

0,01	0,5	38	14,5	MarTest	800 SGA
0,01	0,5	38	36,5	Tesatast	18.10008
0,01	0,5	38	41,2	MarTest	800 SGL
0,01	0,5	40	18,7	Mitutoyo	513-424-10E
0,01	0,5	40	33,3	Mitutoyo	513-414-10E
0,01	0,7	25	4	Starrett	711M
0,01	0,8	28	12,5	B&S BesTest	7030-13
0,01	0,8	28	12,5	Fowler Girod-Tast	52-563-252-0
0,01	0,8	28	12,5	Tesatast	18.10005
0,01	0,8	28	14,5	MarTest	800 S
0,01	0,8	28	17,4	Mitutoyo	513-464-10E
0,01	0,8	35	16	Starrett Swivel	811MPZ
0,01	0,8	35	20	Starrett	709MAZ
0,01	0,8	38	12,5	B&S BesTest	7031-13
0,01	0,8	38	12,5	Fowler Girod-Tast	52-563-452-0
0,01	0,8	38	12,5	Tesatast	18.10006
0,01	0,8	38	14,5	MarTest	800 SG
0,01	0,8	40	17,7	Mitutoyo	513-404-10E
0,01	0,8	40	17,7	Mitutoyo	513-474-10E
0,01	1	35	28	Starrett	709MALZ
0,01	1	38	32,3	MarTest	800 SGB
0,01	1	40	41,0	Mitutoyo	513-415-10E
0,01	1,5	40	18,7	Mitutoyo	513-426-10E

Graduations	Range	Dial Ø	Point L	Brand	Model
0,01	1,6	30	16,5	Interapid (312-2)	74.111367
0,01	1,6	38	16,5	Interapid (312-1)	74.111366
0,01	1,6	38	14,5	MarTest	800 SR

Graduations	Range	Dial Ø	Point L	Brand	Model
0,002	0,2	28	12,5	B&S BesTest	7032-13
0,002	0,2	28	12,5	Fowler Girod-Tast	52-563-253-0
0,002	0,2	28	12,5	Tesatast	18.10009
0,002	0,2	28	14,5	MarTest	800 SM
0,002	0,2	28	11,2	Mitutoyo	513-465-10E
0,002	0,2	35	16	Starrett	708M
0,002	0,2	38	12,5	B&S BesTest	7033-13
0,002	0,2	38	12,5	Fowler Girod-Tast	52-563-453-0
0,002	0,2	38	12,5	Tesatast	18.10010
0,002	0,2	38	14,5	MarTest	800 SGM
0,002	0,2	40	11,2	Mitutoyo	513-405-10E
0,002	0,4	30	15,2	Interapid 312-4	74.111369
0,002	0,4	30	15,2	Interapid 312-3	74.111368
0,002	0,4	38	14,5	MarTest	800 SRM
0,002	0,6	40	11,2	Mitutoyo	513-425-10E

Graduations	Range	Dial Ø	Point L	Brand	Model
0,001	0,14	38	9,2	MarTest	800 SGE
0,001	0,14	40	16,4	Mitutoyo	513-401-10E

VERTICAL METRIC MODELS

Graduations	Range	Dial Ø	Point L	Brand	Model
0,01	0,5	28	36,5	Tesatast	18.10205
0,01	0,5	28	12,5	Tesatast	18.10204
0,01	0,5	28	14,5	MarTest	800 V
0,01	0,8	28	12,5	Fowler Girod-Tast	52-563-272-0
0,01	0,5	38	36	Fowler Girod-Tast	52-563-474-0
0,01	0,8	40	11,2	Mitutoyo	513-455-10E
0,01	0,8	40	17,4	Mitutoyo	513-454-10E
0,01	1,6	30	16,5	Interapid 312-2V	74.111376
0,01	1,6	38	16,5	Interapid 312-1V	74.111375

Graduations	Range	Dial Ø	Point L	Brand	Model
0,02	2	38	36,5	Tesatast	18.10012

Graduations	Range	Dial Ø	Point L	Brand	Model
0,002	0,2	28	12,5	Tesatast	18.10013
0,002	0,2	28	12,5	Fowler Girod-Tast	52-563-273-0
0,002	0,2	38	12,5	Fowler Girod-Tast	52-563-473-0

Manufacturers change models from time to time. Check our website for updates.

CALIBRATION OF GAGES: CALIBRATION PROCEDURES

Often there's no need to send indicators, micrometers, calipers, etc. to gage labs in order to have them calibrated. Gage blocks and standards, on the other hand, must be sent to a lab which specializes in this procedure.

Your gages, instead, can be calibrated in your own shop, and in fact, should be calibrated in your own shop. Ideally, every gage should be calibrated before every use. Only in this way can you be sure that your readings are accurate. Even to comply with various ISO requirements all you need is to label each tool with a serial number and then keep written records of when, where and how often you calibrate them.

In most cases, all you'll need is a set of gage blocks which have been certified. Even inexpensive gage blocks can be used for routine calibration.

Industry standards imply that annual calibration is sufficient for compliance purposes. You should consider the implications of this carefully. You may have damaged your gage after just one use, and then be using an out-of-calibration gage for the rest of the year.

Calibration needs to be customized to the frequency of use of the gage. A gage used once a month can easily be calibrated just once a year. A gage used hourly should probably have a one-month calibration.

Gages in harsh environments need more attention than those used in a cleanroom.

Your quality and production team will have to make the call.

A good way to find out is to start an arbitrary calibration cycle. If everything passes, then you may want to prolong the calibration cycle just to the point where you start to see inaccuracies.

If you're not inclined to calibrate before every use (no one really is) then standard procedure is to calibrate every few months depending on use and wear. If your gage is in constant use then you must choose a more frequent interval.

If you need to follow specific military or industry standards, then you must obtain those standards and by all means, do as they say. Neither the military nor industry always uses common sense in these matters. (Sorry, we can not provide those details. Please confer with the National Institute of Standards [NIST] or an accredited calibration laboratory.)

Herewith some general instructions and guidelines.

TEST INDICATOR CALIBRATION

The fast, economical and accurate way to calibrate a quantity of test indicators is to invest in a Dial Indicator Calibrator with the Test Indicator attachment. These mechanical devices are available in inch or metric models from several manufacturers. They are in effect a micrometer head with a large 3.5" diameter, .00005" accuracy and 0-1" range. The test indicator is positioned above the spindle using the test indicator attachment. The micrometer head is rotated and readings are compared. It will be necessary to have this unit regularly calibrated by a calibration lab to maintain traceability. Ideally, readings should be taken at every numeral printed on the test indicator dial, or as your quality manual requires.

If you need to calibrate large quantities of analog and/or digital indicators you may want to invest in the electronic i-Checker which is hooked up to a computer system and generates inspection certificates. E-mail us for information on this rather costly apparatus ($8900.00 without computer). If you only need to calibrate .001" or .0005" indicators, then you can consider the mechanical indicator calibrator shown on web page 131.

Test indicators can also be calibrated on a surface plate using certified gage blocks. The indicator is securely fastened to a stand and the contact point is brought in contact with a gage block of a given size. The contact point must be parallel with the surface of the block for most manufacturers. Interapid test indicators are an exception and should be at a 12- degree angle, approximately. The gage block can then be removed and replaced a number of times to check for repeatability. Be certain that discrepancies in repeatability are not due to poorly tightened clamps, flimsy stands or other factors. Usually one quarter of a graduation repeatability is allowable, but check with the manufacturer's calibration specs for your particular model.

Errors in repeatability indicate a need for cleaning and, possibly, repair. Do not attempt this without experience.

Accuracy in travel is checked by replacing the gage block with one larger. Very small intervals are required. Ideally you'd want to check the travel at every half revolution, or better. During this procedure be certain that the gage blocks are properly wrung to each other and to the surface. In general, accuracy should not vary more than one graduation per dial revolution on .0005" indicators. Calibration specifications for various manufacturers can be found on this site by referring to the Table of Contents.

If an incremental error occurs - one which increases regularly over the entire travel - then the contact point is of the wrong length or the angle of the point in regard to the surface is incorrect. You should verify that the correct length point is being used. Furthermore, you can make small adjustments by changing the contact point angle. Make repeated calibration attempts with varying angles until you find one which gives correct results. Obviously, it will now be necessary to recreate this same angle when the indicator is used in actual test situations. Some indicators (Girod-Tast, for example) allow you to make small adjustments in length with a set screw.

One final method requires a certified height master. This takes the place of gage blocks. The one we use has an accuracy of .00002". The test indicator is firmly fastened to a test stand and the contact point is positioned (at the proper angle) over one of the height master's test surfaces. Comparison readings are now taken at half-revolution intervals - or better - in both directions.

About the cosine error: for test indicators excluding Interapid models. If the contact point can not be kept parallel to the work surface then you will have to make a mathematical adjustment to the dial reading.

contact point angle	correction factor
10°	reading times 0.98
15°	reading times 0.97
20°	reading times 0.94
30°	reading times 0.87
40°	reading times 0.77
50°	reading times 0.64
60°	reading times 0.50

From this chart you will notice that a contact point held at a 60-degree angle results in one- half the dial reading. Once you determine the angle, simply multiply the dial reading by the corresponding correction factor.

For example, an indicator reading of .0085" at an angle of 30-degrees is equivalent to .0085" x .87 = .0074"

TEST INDICATORS: WHAT CAN GO WRONG?

Starrett Last Word indicator - something went wrong!

Test indicators are also called "dial test indicators" and "lever type" indicators. They're small and the gears, pinions and bearings are even smaller. That's why they're the most fragile of the mechanical instruments, the most easily damaged, and among the hardest to repair. Special tools and considerable practice are usually required. Here are some of the things that are likely to go wrong, how you can prevent it, and what you can do when it happens.

The crystal is the clear plastic lens through which you can see the dial and the hand. Several things can happen. The crystal will become cloudy or discolored with extreme age or solvents you're using in the shop. The solvents could be airborne mist. The crystal can fall out on its own and this happens when the plastic shrinks. This can be due to temperature fluctuations or age of the plastic. The crystal can also fall out when the bezel is damaged. The bezel is the metal or plastic rim around the face of the indicator. In most cases, the plastic crystal is inserted into the bezel with a press. It will snap into place in a groove and because the plastic is now dome shaped it will remain firmly seated. Some manufacturers now use one-piece plastic bezels with crystals. This makes the job of replacing a damaged crystal a snap. Literally. Just snap the new one in place.

However, it also means that you will be able to break the plastic bezel with ease. More on that later.

The crystal can crack from mechanical damage but it can also show stress fractures, tiny cracks which may only show up from certain angles or under certain light conditions. These are most often associated with Compac indicators, even when brand new.

When the crystal is no longer properly seated in the bezel, several things can happen: fluids like oil or coolants will leak under the crystal, may discolor the dial or dissolve the numbers on the face. It will also leak underneath the dial and that's where the hair spring lies. If the hair spring becomes oily, it will no longer function and the indicator won't repeat.

The dials are plastic or painted metal with the numbers and tick marks (graduations) printed on them. Some dials hold up better than others. The numbers can fade with time or because they're washed out by solvents or exposure to extreme light. Dirt and oil can make dials unreadable. If you can remove the bezel and crystal you can use a tissue to wipe the dial, but if you need to replace the dial, then you'll have to remove the hand (or hands) and that needs tools and experience.

The hands (pointers) are preset at different locations on different models. The exact location is really arbitrary and, over the years, various manufacturers have changed their minds on the hand's starting position. Mitutoyo has made it easy —presumably for their own assembly line—by printing a tick mark along the outer edge of the inner dial. When assembling the indicator, the large hand should line up with this tick mark. More significantly, the large hand and the small hand, if there is one, should both coincide to reduce the possibility of reading errors.

Coolants and oils can also leak into the movement from the contact point end. There's no way around this. The body has to have an opening at this point so that the lever can move. Small amounts of fluid may not cause immediate harm, but they will eventually work their way towards the hair spring and then you'll have to have the indicator cleaned. Don't try to immerse the indicator in solvents. It won't work. The indicator has to be disassembled and the hair spring has to be cleaned and dried. Disassembly is complex. Just opening the side of the body won't give you access to the hair spring. You will have to remove the dial assembly and unless you know what you're doing, you'll probably do more harm than good.

In some indicators, the cover on the side of the body can be removed without it affecting the indicator one bit. The Bestest indicators are a prime example. Learning from the watch manufacturers, they've adopted the mono-bloc movement which is independent of the body. In theory, you can open the case, remove the old movement and drop in a new one. In theory. Most other indicators stop working the moment you open the case. And, you'll probably never get them to work again.

The bezel rotates with enough friction to hold it in place during measurements, but with enough ease so that it can be turned while in a set up. The correct balance is tricky and different models use different methods with different results. There is usually nothing you can do if the dial turns too hard or too easily. New plastic bezels of Mitutoyo indicators ride on an o-ring and this works fine while they're new. But, o-rings deteriorate and solvents will ruin their elasticity. The bezel becomes very hard to turn or it practically falls off on its own. Swiss made indicators use a metal spring which could be adjusted, but it requires disassembly. If you have special requirements, ask your repair shop to make the modifications for you.

The bezel can be bent out of round in any number of ways. I'm sure you can think of some. This will cause the crystal to pop out. Sometimes the bezel can be bent back in shape, but often you'll have to replace it. Bezels can be bought with crystals already installed which is a good thing if you don't have access to a crystal press. The bezels of Bestest indicators can not simply be replaced. They require instrument disassembly. Bestest bezels are quite thin walled and will easily be bent out of shape. Compac and Interapid have more substantial bezels which can take a bit of abuse. Forget about the plastic bezels of Mitutoyo. They'll crack for any number of mysterious reasons.

The contact points on test indicators are model specific. Each model, each manufacturer, has a specific length of point and this is important since the measurement depends on a leverage system. The length of the points from the center of the ball to the center of the pivot enters into the calculation. Information on contact points can be found on web page 21.

The contact point swivels on all indicators. They're held in place by a friction clutch in most cases. The Last Word indicator and Gem indicator have a ratchet instead. You can move the point almost 180 degrees in some cases. Do this with your thumb and finger even if the friction seems very stiff. Compac and Interapid indicators are notoriously stiff. It's designed that way so that the point won't move on its own when you are measuring. Do not use pliers. It's not necessary and it will probably cause harm.

To obtain maximum repeatability it will be necessary to mount your indicator firmly. Nothing must move except the contact point. Mounting is done using the dovetails or mounting stems which most indicators have. A few (Starrett Last Word) don't have dovetails and you'll have to use special holders. Interapid indicators have mounting stems attached to the body as well as dovetails but even here the manufacturer suggests mounting by the dovetails for maximum stability. The amount of friction on the Interapid stem can be adjusted by tightening the oversized mounting screw. If these attachments are not tight, or if the fixture you're using (perhaps a magnetic base) is flimsy, then you'll have problems with repeatability.

Dovetails are (or should be) of uniform dimensions from one manufacturer to the next, making it possible to mount any indicator on any fixture. We've noticed some Chinese indicators and mag bases with tolerances so lax that they don't easily fit.

Other repeatability problems can be traced to the contact point. It may not be screwed in tight enough. Or, you may feel sideways play in the pivot. In quality indicators, the pivot rests in two ball bearings which may be adjustable. If there is sideways play, and the bearings are not damaged, then they can be carefully tightened. Special tools are used to accomplish this task. Of course, it's possible that the ball bearings are damaged, and you may not see this from the outside. Most ball bearings are small and hitting the contact point (or dropping it on the floor) will damage the bearings. Nothing can be done but to replace them. It's not an easy thing to do. It actually requires complete instrument disassembly which you know if you've ever tried it. Compac indicators have oversized ball bearings which almost never need replacing. Bestest bearings are very susceptible to damage.

The indicator hand may not repeat if the hair spring is bent out of shape or has gotten oily (as mentioned before). This requires professional help. In some cases, an indicator can become magnetized and this will affect repeatability. Run it through a demagnetizer to see if it helps. Starrett Last Word indicators can suffer from this malady. Mitutoyo indicators are "antimagnetic."

The indicator will be sluggish if the ball bearings have become gummed up, or if they're too tight, or if the hair spring is oily. Amateur repairmen will usually over-tighten the bearings, thus ruining them. In other cases, cleaning or bearing replacement is necessary.

If the indicator hand hangs up at one spot, or if it appears to jump, then there may be a damaged gear tooth. Several small gears are used in translating the movement of the contact point into the movement of the hand. These gears are small and the teeth are easily bent out of shape due to shock. It could also be the case that a bit of dirt, a metal chip, some dried grease has gotten into the teeth. Or, take a close look, the large hand may be touching the dial or the crystal or, if it has two hands, one might be touching the other. You may be able to bend the hand but you'll probably break the center pinion in the process. Be warned.

In severe cases, the pinions on the gears break off or become corroded. Corrosion will happen in all indicators exposed long enough to water or humidity. It's called rust. Rust usually makes an indicator not work. It will probably be necessary to replace many of the internal parts when this happens. If your indicators continually become rusty, and you can't change the environment, then consider switching to another brand of indicator. Although we don't know this for sure, some brands may be less susceptible to rust than others.

The Long Island Indicator website (www.longislandindicator.com) contains many other web pages with information on test indicators. Please refer to our home web page for subjects.

Indicator Crystal Installation Reference Page

The crystals (clear plastic lens) for test indicators, dial indicators and calipers are typically larger than the bezel for which they are designed.

New crystals are also flat. They become domed once they are inserted, under pressure. An old crystal will appear domed because, after a long time, it will take on this shape. Unfortunately, at this point it will also fall out easily.

The crystals have a beveled edge which will fit into the groove on the upper part of the bezel, but must be inserted in a concave (domed) manner so that the crystal will remain in place under pressure, and so that the plastic clears the movable dial hand. You will need to order a crystal that is at least 1 mm larger in diameter than the bezel requires.

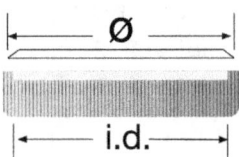

The flat crystal has to be larger than the inside diameter of the bezel (where the crystal will fit).

A crystal press is usually required to give the plastic lens its concave shape. (Bestest indicators and B&S calipers are exceptions.) Place the crystal and bezel in the press, in the correct orientation (take into consideration the beveled edge: the smaller surface will become the upper surface) and while the press is bending the plastic, gently snap the bezel into place. Release the press and the crystal will be firmly seated and have the correct curvature.

The press can not be used with crystals made of glass, which will shatter under pressure.

A Crystal Press is shown with various inserts. The photographer, not knowing differently, mounted the lower cup incorrectly. It should be the other way around. Refer to the video below.

This hand operated press designed for use with Mitutoyo indicators, will allow you to insert crystals into bezels on test indicators, dial calipers and dial indicators up to 2-1/4" diameter. Comes with 8 different inserts as shown above. It can also be used for other brands of indicators including Interapid, Compac and Starrett. It is not suitable for Bestest indicators, Brown & Sharpe calipers or Federal dial indicators. Make sure that this press will work for the gages you have by reading the instructions printed below.

➡ Hint: Use a bottom cup the same size or one size smaller than the diameter of the crystal.

Glue or cement is never used to hold the crystal in place. Some manufacturers may use a thin bead of silicon to create a waterproof seal. This is a tricky maneuver and should only be attempted with caution.

While home-made presses can be improvised, for frequent use and reliable results we suggest ordering the portable Crystal Press #7000 shown above. It allows for easy crystal replacement on all size test indicators (except for BesTest and TesaTast) and AGD series 1 and 2 dial indicators (Federal indicators excepted), and certain dial calipers. Before buying the press, read the instructions below to make sure a press will work with your indicators or calipers.

CRYSTAL SIZE CHART

Even though the manufacturer's name may be listed next to a particular crystal in the ordering chart below, it does not mean that every indicator or caliper made by that manufacturer can use that crystal. Be sure to measure the inside diameter of the bezel. (Read instructions above carefully)

Note: this is the maximum diameter (±0.1 mm) of the plastic crystal when flat. After the crystal is installed, it will be domed. If you measure an old crystal, which is already domed, take this into consideration.

Diameter	Suitable for (unless otherwise noted)
25.1 mm	Alina, Compac (small dial models)
26.7 mm	Standard Gage, Mitutoyo
27.9 mm	BesTest (old models), Girod, Puppitast, TesaTast
28.6 mm	MarTest, China, Interapid, Federal "A", B&S Group O
30.8 mm	Mitutoyo "Pocket" style, Starrett models made in China
34.7 mm	Mitutoyo, Teclock
35.7 mm (special .032" thickness)	Brown & Sharpe, TESA, Etalon, Compac
36.6 mm	Starrett model 196, Interapid
36.8 mm	Starrett
37.8 mm	Girodtast, Tesatast, Mitutoyo, B&S, Starrett
39.0 mm	Alina, Puppitast, MarTest, Compac, Boice
40.0 mm	Starrett
42.0 mm	Mitutoyo
52.1 mm	Peacock, Intertest
52.1 mm (special)	Compac dial bore gage (old style)
53.2 mm	Boice, Starrett, Mahr-Federal
53.8 mm	CDI, Mitutoyo, S-T, Standard, Starrett, Teclock

How to Replace a Test Indicator Crystal

We list the indicators here by their manufacturers' names. Many indicators are sold with vanity names, so you may have to find out who the real manufacturer is. Order the appropriate sizes from the list above.

Alina: loosen, but don't remove, the three tiny screws under the bezel which can now be pried off, with your hands, like the lid of a take-away coffee cup. You may have to twist and turn the bezel a bit until it works. They can occasionally be stubborn. You'll need a press to insert the 25.1 mm diameter crystal (small) or the 39.0 mm diameter crystal (large).

BesTest (current models): the old crystals are removed with a simple tool that acts like a suction cup. You can improvise other methods but you're likely to cause damage. If the bezel is still round and in very good condition, you can probably just press the new crystal in place using your thumbs. If the bezel is dented or bent out of shape, then it's best to leave this to someone with the right tools. See online parts list.

Bestest (old models): the bezel can be pried off with a flat bladed screw driver. A metal spacing ring is removed from behind and the old crystal is pushed out with your thumb. Careful you don't cut yourself on the metal bezel. The new crystal is pushed in from below, again with your thumb. It is slightly over sized so that it will become convex in the process. The metal retainer is also put back in and the bezel is pressed back in place on the indicator. Use 37.8 mm diameter crystal (for the large bezel) and #240 (for the small 28mm bezel). The much smaller 25 mm bezel, rarely seen, uses a crystal which we no longer carry.

- Brown & Sharpe 7030-2 uses crystal #240 (press not required)
- Brown & Sharpe 7038-2 uses 37.8 mm Ø crystal (press not required)

Brown & Sharpe / Bestest briefly had its name on what is an Interapid Swiss-made indicator. See instructions for Interapid.

- Brown & Sharpe 7028-4 uses the 36.6 mm Ø crystal

Compac: remove the spring which is visible in the groove on the underside of the bezel. Use a thin needle to get the spring out. Try not to bend this spring out of shape. The bezel, if undamaged, lifts off easily. There is no need to remove anything else. You'll need to use a crystal press to get the new crystal into the bezel. Original Swiss crystals are brittle. If you apply too much pressure with the press, they'll shatter. When assembling make sure the tab on the outer dial fits into the space allotted for it on the inside of the bezel. If the bezel now turns too easily or too hard you can make adjustments by reshaping the spring.

- Small bezels: 25.1 mm Ø crystal
- Large bezel: (models with the letter G) 39.0 mm Ø crystal

Important note: do not attempt to remove the bezels on the Compac 240 series test indicator (which has one revolution). Doing so, will result in the loss of hair spring tension. Unless you know what you are doing, send these to a repair shop for servicing.

Federal Testmaster: The old style Testmaster indicator has a bezel which can be tricky to remove. A slot on the side of the bezel will allow you to depress the spring which holds the bezel in place. Rotate the bezel until the spring is visible and then use a thin bladed screw driver to press the spring. If you can't get this to work and resort to prying the bezel off, you will inflict some damage on the top plate. The small bezel (about 1" diameter) takes crystal #210 which must be inserted with a press.

Gem test indicator: very much like the Starrett Last Word, the bezel is pried off with a screw driver. You can discard the wire ring and install the Starrett Last Word crystal. These crystals are molded to shape so you won't be able to use any of the flat crystals listed on this page.

Girod: the thin-walled bezel can be carefully pried off with a screwdriver. The new 37.8 mm diameter crystal (for the large bezel) and #240 (for 28mm bezel) can be pressed into the bezel from below using your thumbs. Don't forget to put the spacing ring back before pressing the bezel onto the body.

Interapid: the bezel is threaded and unscrews. Hold the lower bezel plate steady with a small screw driver placed in one of the two holes while you unscrew the bezel. See online parts list.

- Interapid 312B-3 uses the 36.6 mm Ø crystal (crystal press required)
- Interapid 312B-2V uses the 28.6 mm Ø crystal (crystal press required)

Kafer: once you (carefully) pry off the retaining ring on top, the crystal just drops out. Replace it and you're set to go. Kafer crystals are offered elsewhere on this web site.

- 37.3 mm Kafer 5.2109 (plastic without beveled edge, lies flat)

- 53.8 mm Kafer 5.2101 (plastic 1.1 mm thick without beveled edge)

Mahr: Puppitast has a thin-walled bezel which can be pried off with a screwdriver, but do this carefully. The new 27.9 mm diameter crystal (for the small faced models) can be pressed into the bezel from below. No special equipment is needed. Remember to put the spacing ring back into place before pressing the bezel back onto the indicator. The larger diameter bezels (about 1-5/8 inch) behave the same way and require the 39.0 mm diameter crystal.

Puppitast 800 SM uses the 39.0 mm diameter crystal The oldest model Puppitast lift off the same way but they do not have a spacing ring and you will need a crystal press to insert the 39.0 mm diameter crystal into the 40 mm diameter bezels.

Mahr-Federal new style indicators "MarTest" have bezels which ride on a rubber o-ring. They can be pried off but you'll need a press to insert the 39.0 mm diameter crystal into the 1.5" diameter bezel. The smaller 30 mm bezel will take the 28.6 mm diameter crystal.

Mitutoyo: newer models have plastic bezel & crystal combinations which snap on and off. Nothing could be easier and this is an excellent selling point.

Older models have bezels which can be pried off with a small screw driver. For the new series 513-412 and similar models (the combination black plastic bezel and crystal) you may also have to replace the rubber o-ring if the old one has stretched out of shape. The fit should be snug. Some Quick-Set and "Pocket" models still have metal bezels and a crystal press will be required to insert the crystal. See page 162 for a complete parts list.

- 513-462 and 513-463 uses crystal ø 26.7 mm (crystal press required)
- 513-518 and other "Pocket" style indicators use the 30.8 mm Ø crystal (crystal press required)

Peacock: (Pic Test) a wire spring holds the 34 mm diameter bezel in place. Three very small screws hold the 47 mm bezel in place. Don't lose these screws! For the 47 mm bezel, use crystal #520 which requires a press for insertion.

SPI: some of these models are made in China. Usually, when there is no country of origin printed on the dial, it's Chinese. For the new Interapid look-alike beginning with the number 14, you will be able to pry off the bezel with a screw driver and, after removing the metal spacing ring, you can press the new crystal in place from behind, using your thumb. The 36.6 mm diameter crystal will do the trick. If the SPI indicator is made in Japan, then you'll want to refer to Teclock below. If it is made in Switzerland, then you'll want to see Compac above. If you're not sure, just send us the indicator with "replace crystal only" instructions.

Starrett Last Word: remove the chrome bezel by prying it off with a screw driver and insert the crystal from below (part PT07112) into the bezel. The crystal fits easily without resorting to a press or any other equipment. If your indicator still has a wire spacer, throw the wire away. The new crystals don't need this spacer. A pair of jeweler's pliers will help you squeeze the bezel back into place. This part can be ordered directly from the manufacturer. The discontinued model 711-T1 used a crystal that had to be inserted with a press.

Starrett: the bezels for new models 708 and 709 series are held on by an o-ring. Use a screwdriver to wedge the bezel off. You'll need a crystal press to insert the crystal with 36.8 mm diameter. Snap the bezel back on. (The original Starrett part number is PT19043 if you would rather buy this part from a Starrett distributor).

Starrett models made in China: pry the bezel off with the blade of a screwdriver. No tools are needed to replace the crystal. The old one can be pushed out and the new one (30.8 mm) fits snuggly inside. When you press the bezel back onto the indicator, the crystal will also be pushed into proper alignment.

Teclock: assuming the dial diameter on this test indicator is about 1-3/8" then you'll want crystal #S035691. You'll find a retaining spring on the underside of the bezel which has to pried out with a pointy tool. When the bezel is off, you can pop out the old crystal using your thumbs (careful you don't cut yourself). The new crystal can be pressed in place with your fingers if the bezel is still round and in good condition. Otherwise, a crystal press will help insert it. The flat 34.7 mm diameter crystal will also work but requires a press for certain.

Tesatast: (old models without model numbers) the thin-walled bezel can be carefully pried off with a screwdriver. The new 37.8 mm diameter crystal (for the large bezel) and 27.9 mm diameter (for 28mm bezel) can be pressed into the bezel from below using your thumbs. Don't forget to put the spacing ring back before pressing the bezel onto the body. For new models, see Bestest above.

HOW TO READ THE GRADUATIONS ON INDICATOR DIALS

As a standard convention, every number printed on the dial of an inch-reading test indicator or dial indicator represents .001" This applies to all makes and models which are inch reading. In fact, it also applies to dials on thickness gages, dial calipers, dial bore gages, and dial micrometers.

- Whether .001", .0005", or .0001" the same rule applies. Inch reading dials will be white, black and sometimes other colors (but not yellow which is reserved for metric dials)

HOW TO READ THE GRADUATIONS ON METRIC DIALS

The same principle applies to all metric dials. In this case, each number will represent 0.01 mm regardless of the gage's discrimination. For example the number 10 would represent 0.10 mm and the number 90 would represent 0.90 mm.

- By convention, metric indicators sold in the U.S.A. should have yellow dials. Metric dials in the rest of the world are white.

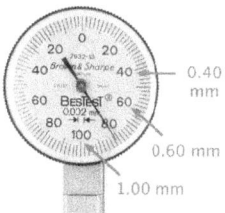

www.longislandindicator.com

BROWN & SHARPE BESTEST INDICATOR REPAIR MANUAL

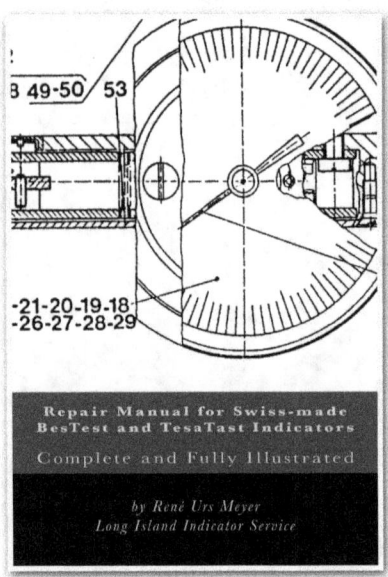

Repair Manual
for BesTest Indicators
By René Urs Meyer
Is only available at Amazon.com

Table of Contents
Specifications ... 6
Can it be fixed? ... 7
Repair Tools ... 10
Disassembly ... 14
Cleaning ... 23
Reassembly ... 27
Congratulations! ... 58
Calibration ... 59
BesTest Contact Points ... 61
TesaTast Contact Points ... 64
BesTest Dials ... 66
Horizontal test indicator spare parts list ... 68
Vertical test indicator spare parts list ... 71
Illustrated Spare Parts ... 75

www.longislandindicator.com

LAST WORD INDICATOR REPAIR MANUAL

Last Word Indicator
Repair Manual
by René Urs Meyer is only
available at Amazon.com

Table of Contents

Introduction ... 4
Tools you will need ... 5
Can it be fixed? ... 6
Disassembly ... 7
How to remove the bezel and crystal ... 9
Cleaning ... 16
Reassembly ... 19
Replacing a broken jewel ... 25
Replacing the hair spring ... 32
When there is no clearance in the spiral pinion ... 38
Installing the lever assembly ... 40
When the pivot screw is loose ... 42
When the contact point is loose ... 43
Checking the engagement of the lever assembly ... 45
What happens if the jewel is too far down ... 45
Setting the hair spring tension ... 46
Reshaping a damaged dial ... 48
Repairing a damaged bezel ... 53
Congratulations! Your repair is complete ... 59
What does it cost? ... 59
Calibration ... 60
The Professional Touch ... 61
Spare parts breakdown ... 62
Illustrated Spare Parts ... 65
Other books which you will find useful ... 77

www.longislandindicator.com

INTERAPID INDICATOR REPAIR MANUAL

Table of Contents

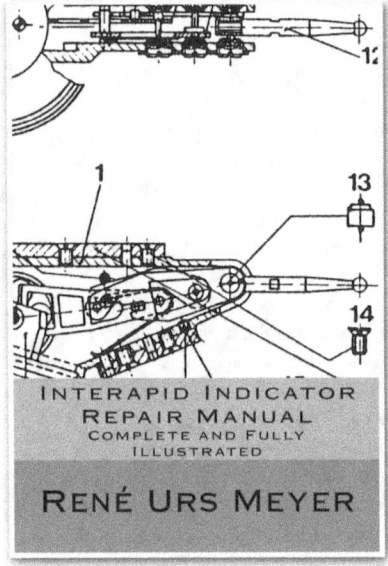

Interapid Repair Manual is only available at Amazon.com

Tools you will need ... 6
Can it be repaired? ... 8
Disassembly ... 10
Cleaning ... 21
Basic Reassembly when nothing is broken... 25
Test the indicator for functionality and calibrate it ... .53
Special repair situations ... 55
Replacing the 4 mm stem attachment ... 56
How to repair a damaged bezel ... 61
How to remove a bezel when it refuses to unscrew ... 64
How to repair the contact point pivot assembly ... 67
When the dial screws come loose ... 72
Interapid 312B-15 and 312B-15V considerations ... 73
When the front end is damaged ... 74
Damaged dove tails ... 76
Repairing damaged movement bearings ... 77
Repairing a damaged crown gear ... .80
Adjusting the crown gear bearings ... 81
Replacing the crown gear bearings ... 83
When the entire indicator movement has broken off ... 84
Repairing the hair spring ... 85
Replacing damaged ball bearings ... 91
Parts breakdown .0005" and .001" (0,01 mm) ... 95
Parts breakdown .0001" (0,002 mm) ... 96
Movement breakdown (all models) ... 97
Parts List for Interapid Series 312 ... 98
Contact Points ... 102
Questions and Answers ... 105

www.longislandindicator.com

For additional and up-to-date information on

repairs
test indicators
dial indicators

dial bore gages
micrometers
indicating micrometers

and other essential measuring tools

please visit our web site
www.longislandindicator.com

www.ingramcontent.com/pod-product-compliance
Lightning Source LLC
Chambersburg PA
CBHW081431220526
45466CB00008B/2350